AMINO ACIDS AND PROTEINS IN FOSSIL BIOMINERALS

New Analytical Methods in Earth and Environmental Science Series

Because of the plethora of analytical techniques now available, and the acceleration of technological advance, many earth scientists find it difficult to know where to turn for reliable information on the latest tools at their disposal, and may lack the expertise to assess the relative strengths or potential limitations of a particular technique. This new series will address these difficulties, and by providing comprehensive and up-to-date coverage, will rapidly become established as a trusted resource for researchers, advanced students and applied earth scientists wishing to familiarize themselves with emerging techniques in their field.

Volumes in the series will deal with:

- the elucidation and evaluation of new analytical, numerical modelling, imaging or measurement tools/techniques that are expected to have, or are already having, a major impact on the subject;
- new applications of established techniques; and
- interdisciplinary applications using novel combinations of techniques.

See below for our full list of books from the series:

Boron Proxies in Paleoceanography and Paleoclimatology
Barbel Honisch, Stephen Eggins, Laura Haynes, Katherine Allen, Kate Holland, Katja Lorbacher

Structure from Motion in the Geosciences
Jonathan L. Carrivick, Mark W. Smith, Duncan J. Quincey

Ground-penetrating Radar for Geoarchaeology
Lawrence B. Conyers

Rock Magnetic Cyclostratigraphy
Kenneth P. Kodama, Linda A. Hinnov

Techniques for Virtual Palaeontology
Mark Sutton, Imran Rahman, Russell Garwood

AMINO ACIDS AND PROTEINS IN FOSSIL BIOMINERALS

An Introduction for Archaeologists and Palaeontologists

BEATRICE DEMARCHI

Department of Life Sciences and Systems Biology,
University of Turin, Italy

WILEY Blackwell

This edition first published 2020
© 2020 John Wiley & Sons Ltd

Registered Office(s)
John Wiley & Sons, Inc., 111 River Street, Hoboken, NJ 07030, USA
John Wiley & Sons Ltd, The Atrium, Southern Gate, Chichester, West Sussex, PO19 8SQ, UK

Editorial Office
9600 Garsington Road, Oxford, OX4 2DQ, UK

For details of our global editorial offices, customer services, and more information about
Wiley products visit us at www.wiley.com.

Wiley also publishes its books in a variety of electronic formats and by print-on-demand.
Some content that appears in standard print versions of this book may not be available in
other formats.

Library of Congress Cataloging-in-Publication data applied for
ISBN: 9781119089445 [hardback]

Cover Design: Wiley
Cover Image: © oxygen/Getty Images

Set in 10/12pt Minion by SPi Global, Chennai, India

Printed and bound by CPI Group (UK) Ltd, Croydon, CR0 4YY

10 9 8 7 6 5 4 3 2 1

Contents

Preface   vii
Acknowledgements   ix

1   Biominerals and the Fossil Record   1
    Why Study Old Biominerals?   1
    What are Biominerals?   4
    How and Why are Biominerals Formed?   5
    'Biomineralization Toolkit': From Proteins to Proteomes   8
    Fossil Biominerals, Fossil Proteomes   12
    References   16

2   Mechanisms of Degradation and Survival   23
    Introduction   23
    Hydrolysis   27
    Racemization   31
    Decomposition and Other Diagenesis-induced Modifications   35
    References   38

3   Proteins in Fossil Biominerals   43
    Bone and Other Collagen-based Hard Tissues   43
    Tooth   45
    Eggshell   47
    Mollusc Shell   51
    Other Substrates   57
    References   63

4   Chiral Amino Acids: Geochronology and Other Applications   71
    Dating the Quaternary (Pleistocene and Holocene)   71
    Principles of AAR Dating   76
    Measuring D/L Values   78
    Factors Affecting D/L Values   82
    Aminostratigraphy   87
    Aminochronology   92
    Palaeothermometry   97
    Testing the Suitability of Biominerals for Geochemical Analyses   98
    Taxonomic Identification   99
    Appendix: Practical Tips on How to Plan and Conduct
       an AAR Study   100
    References   104

5   Ancient Protein Sequences   113
    Ancient Protein Analysis by Mass Spectrometry   113
    Ancient Proteins: Past and Future   120
    References   122

Index   127

Preface

This book is largely concerned with biominerals. In particular, it introduces the proteins responsible for biomineral formation and discusses the information these proteins may yield when surviving in the archaeological and palaeontological records.

The historical perspective on the study of proteins in biominerals (from the discovery of fragments of molecules entombed in fossils to the huge advances in shotgun proteomics that are currently underway) offers an exemplary tale of a research landscape that is becoming increasingly and essentially inter- and trans-disciplinary. Understanding the patterns of survival and degradation of biomolecules in the biogeosphere is the focus of palaeobiogeochemistry. However, the information we can gather from the extent of preservation or breakdown of these organics spans at least three fields of research: biomaterials (engineering), biochemistry and geochemistry (including geochronology and applications to archaeological, palaeontological, geological and palaeoclimatic questions), and evolutionary biology.

The author's own experience followed a trajectory beginning with cultural heritage and archaeological sciences, dipping into geology and analytical chemistry, then attempting to bring together new research in biomineralization and ancient proteins in order to elucidate mechanisms of protein survival. This volume reflects the cross-disciplinary interest of the author and aims to become a useful reference for postgraduate students and researchers embarking on a career as a biomolecular archaeologist or palaeontologist.

Chapter 1 briefly highlights the reasons for which studies in biomineralization are fundamental to understanding biomolecular survival in the fossil record, while Chapter 2 describes the main mechanisms of protein degradation. The composition of a range of biomineralized proteomes is summarized in Chapter 3, followed by a discussion on the two main applications of ancient protein studies, i.e. amino acid racemization/protein diagenesis dating (Chapter 4) and palaeoproteomics (Chapter 5).

Acknowledgements

I am deeply grateful to Matthew Collins, Darrell Kaufman, Frédéric Marin, Kirsty Penkman and John Wehmiller for their valuable comments on the text. All remaining mistakes are, of course, mine alone. Kirsty and Matthew, in particular, have been an immense source of scientific and personal inspiration over the years.

I also wish to acknowledge my colleagues and friends at BioArCh, University of York (UK), with whom I shared a vision of interdisciplinarity, and those at the University of Turin (Italy), who are supportive of my work.

The editors at Wiley have been a model of forbearance and I thank them for this.

I am incredibly fortunate for having the continuous and unwavering support of my family, for which I cannot thank them enough.

1 Biominerals and the Fossil Record

1.1 Why Study Old Biominerals?

Biomineralized tissues have a good chance of surviving in the fossil record and of preserving, like a geochemical time capsule, a snapshot of the environmental conditions in which the organism (or tissue) itself was formed. This has been known since the first decades of the 20th century and is the basis for reconstructing past variations in sea temperature by measuring the stable isotopic composition of fossil biominerals (e.g. shells, foraminifera; Emiliani, 1955). Studying the way in which biominerals are formed can thus reveal important information about our planet's past. At the same time, it can teach us how to build new materials with superior characteristics (so-called biomimetic and bioinspired materials).

However, fossil biominerals can also reveal the evolutionary history of life itself. Heinz Lowenstam, an intellectual giant of biomineralization, presented his famous table of biomineralized organisms (redrawn here as Table 1.1) at the 'Biogeochemistry of Amino Acids' meeting at Airlie House, Warrenton, Virginia, in 1978 (Lowenstam, 1980). Observing that the evolution of biomineralized organisms has its roots in deep time (~650–550 Ma ago), and that the majority of biominerals 'appear' during the Cambrian 'explosion', he suggested that: 'extended to the fossil record of the Phanerozoic, [...] the study of the organic and bioorganic fractions [in biominerals] holds promise to trace pathways of biochemical evolution.' Lowenstam's proposal was not without substantiation: in 1978 it had already been known for a couple of decades that organic matter could be recovered from fossils (Abelson, 1954). This is due to the fact that the mineral component can protect the biological fraction (*biomolecules*) from degradation: after the death of an organism, a range of biological and chemical factors immediately induce the breakdown

Amino acids and Proteins in Fossil Biominerals: An Introduction for Archaeologists and Palaeontologists, First Edition. Beatrice Demarchi.
© 2020 John Wiley & Sons Ltd. Published 2020 by John Wiley & Sons Ltd.

Table 1.1 Types of biogenic minerals and their occurrence in extant phyla, redrawn from Lowenstam (1980).

Mineral group	Mineral	MONERA	DINOFLAGELLATA	HAPTOPHYTA	BACILLARIOPHYTA	PHAETOPHYTA	RHODOPHYTA	CHLOROPHYTA	ZYGNEMATOPHYTA	RHYZOPODEA	SYPHONOPHYTA	CHAROPHYTA	HELIOZOATA	RADIOLARIATA	FORAMINIFERA	MYXOMYCOTA	CILIOPHORA	BASIDIOMYCOTA	DEUTEROMYCOTA	PORIFERA	COELENTERATA	PLATYHELMINTHES	ECTOPROCTA	BRACHIOPODA	ANELLIDA	MOLLUSCA	ARTHROPODA	SIPUNCULA	ECHINODERMATA	CHORDATA	BRYOPHYTA	TRACHAEPHYTA
		MONERA	**PROTOCTISTA**															**FUNGI**		**ANIMALIA**											**PLANTAE**	
HALIDES	Fluorite																										*	*	*			
HALIDES	Amorph. Fluorite Precursor																										*		*			
OXALATES	Whewellite						?	*		*								*	?								*				*	*
OXALATES	Weddelite																	*	?								*		*	*		
SULFATES	Gypsum																									*					?	*
SULFATES	Celestite													*																		
SULFATES	Barite											*	*																			
SILICA	Opal				*			?					*	*	*					*					*	*	*		*			*
Fe-OXIDES	Magnetite	*																								*	*		*			
Fe-OXIDES	Maghemite	?																														
Fe-OXIDES	Goethite																									*						
Fe-OXIDES	Lepidocrocite																		*							*						
Fe-OXIDES	Ferrihydrite	*																*	*							*	*		*		*	*
Fe-OXIDES	Amorph. Ferrihydr.																*									*			*			
Mn-OXIDES	"Todokorite"	*																														
Fe-SULFIDES	Pyrite	*																														
Fe-SULFIDES	Hydrotroilite	*																														
CARBONATES	Calcite	*	*	*			*				*		*	*	*			*	*	*	*	*	*	*	*	*	*	*	*	*	*	*
CARBONATES	Aragonite	*	?		*	*	*			*			*					*	*	*	*			*	*	*	*	*	*	*		*
CARBONATES	Vaterite						*																			*	*		*	*		*
CARBONATES	Monohydrocalcite	*																								*			*			
CARBONATES	Amorphous Ca Carbonate																*			*						*	*					
PHOSPHATES	Dahllite	*																*	*										*			?
PHOSPHATES	Francolite																							*	*	*			*			
PHOSPHATES	$Ca_3Mg_3(PO_4)_4$																									*						
PHOSPHATES	Brushite																									*						
PHOSPHATES	Amorph. Dahllite Precursor																		*							*	*		*			
PHOSPHATES	Amorph. Brushite Precursor																									*						
PHOSPHATES	Amorph. Whitlockite Precursor																									*	*		*			
PHOSPHATES	Amorph. Hydr. Ferric Phosphate																									*	*		*			

of the organic matter (soft tissues), but if these organics are somehow protected by the mineral phase, then degradation will be slow (or slower). It follows that ancient molecules (nucleic acids, proteins, lipids, carbohydrates) will be found mainly in sub-fossil (hereafter "fossil") mineralized substrates, and that we can target these for molecular palaeontology/archaeology.

At the time of writing, the field of palaeobiogeochemistry has expanded so much that we now use 'fossil' biomolecules routinely in order to recover information on the age since death of an organism (e.g. in the case of protein diagenesis/amino acid racemization dating; see Chapter 4), as well as on the physiology, phylogeny, and biogeography of extinct and extant organisms (from Neanderthals to crops), the evolution of diet, agricultural and husbandry practices, and even diseases. Indeed, we now have a whole dedicated field of 'biomolecular archaeology', which can be split into different subdisciplines (palaeogenomics, palaeoproteomics, palaeolipidomics, stable isotope geochemistry). Many recent reviews of the literature deal with the potential of new technologies for studying ancient biomolecules (Cappellini et al., 2018; Hendy et al., 2018; Welker, 2018).

And yet, these technological developments have not so far been able to fulfil Lowenstam's dream: in order to reconstruct evolutionary patterns in deep time, it is necessary to retrieve sequence data (DNA or proteins, bearing phylogenetic information), from *really* old fossils. Since DNA degrades more rapidly than proteins, the latter would be the molecules of choice. Despite repeated claims of preservation of proteins in Cretaceous (dinosaur) samples (Abelson, 1956; Miller and Wyckoff, 1968; Schweitzer et al., 1997; Asara et al., 2007; Cleland et al., 2015; Schroeter et al., 2017), which have been repeatedly dismissed (Collins et al., 2000; Buckley et al., 2008, 2017; Saitta et al., 2019), the oldest endogenous peptide sequences have been reported from Plio-Pleistocene fossils (eggshell: Demarchi et al., 2016; tooth enamel: Cappellini et al., 2019). In fact, comparatively little effort has gone into systematically assessing the survival potential of organic matrices other than those for bone, eggshell and teeth. (Incidentally, bone has been taking the lion's share of the effort, despite its being well known as a poor repository of endogenous molecules due to its porous and 'leaky' nature.) As a result, many scientific papers reporting protein sequence data bear two striking features: (1) the missed opportunity of linking research in ancient biomolecules with the latest theories and discoveries in the field of biomineralization; (2) the limited engagement with issues of diagenesis, which were instead at the forefront of debate at the time Lowenstam published his work.

In order to fill this gap, this book will focus on the mechanisms of degradation and survival of proteins in fossil biominerals, and look at the extent to which this affects the use of ancient proteins as a means of understanding the past. Applications for dating purposes, as well as in the newer field of 'palaeoproteomics', will be considered. But, before we begin, we must clarify what biominerals actually *are*.

1.2   What are Biominerals?

Biominerals are hybrid nanocomposite biomaterials, in which an organic fraction produces mineralized structures (skeletons) from inorganic precursor ions (Weber and Pokroy, 2015). Or, as Weiner and Dove define them, 'where the distinctions between the chemical, biological and earth science disciplines melt away' (Weiner and Dove, 2003). Biominerals display highly sophisticated architectures, exceptional mechanical properties and serve a variety of purposes, from protection of the soft body to scattering of light (Addadi and Weiner, 2014). Among the many studies which review research on biominerals, a recent historical overview has been provided by Addadi and Weiner (2014). This, like many others, highlights the fact that, despite the natural variety of biogenic products, their *chemical composition* is generally very similar: two-thirds of known biominerals are made of calcium carbonate (in its various polymorphs, i.e. calcite, aragonite, vaterite and, in some cases, amorphous) or calcium phosphate (e.g. bone apatite). Silica-based skeletons are also formed by some organisms, while other minerals, e.g. magnetite, are rarely found in biominerals (Lowenstam, 1981; Carter, 1989).

The organic fraction of biominerals is highly complex and contains, among other things, water, which is mainly present as fluid inclusions in the carbonate (Hudson, 1967; Gaffey, 1988). In shells, the amount of water has been found to be consistent within the same species but extremely variable between different taxa, most likely because these inclusion waters are vestiges of metabolic fluids produced by the tissues (Lécuyer and O'Neil, 1994). Other than water, the organic compounds present in biominerals are lipids (n-alkanes, free and bound cholesterol, and bound fatty acids: Stern, 1996), carbohydrates and proteins. Proteins represent the biggest fraction (~65–90%), followed by lipids (0.8–3%) and carbohydrates (0.2–0.3%), according to the estimates of Goulletquer and Wolowicz (1989). However, their study only compared three species of mollusc shell and it is likely that these percentages vary considerably, especially for carbohydrates: for example, chitin, one of the main carbohydrates in many biominerals, can represent up to 5–10% of the organic matrix. It is noteworthy that water is integral to the structure of organic molecules (up to 20% by weight in proteins), and that, as such, this hydration layer plays an important role in both the functioning and the degradation of the molecules themselves (see Chapter 3). In general, there is a striking lack of data on the composition of the organic fraction in different biominerals. Despite this poor knowledge, and notwithstanding the fact that it is commonly accepted that the three classes of biomolecules must all play a role in biomineralization (Farre and Dauphin, 2009; Luquet et al., 2012; Rao et al., 2014), it is mainly the protein fraction that has been studied.

The way in which inorganic and organic matrices are organized spatially in different biominerals is also still far from clear, despite the fact that the organic matrix within biogenic minerals was first observed around 50 years ago, using transmission electron microscopy (TEM). These early studies of the bivalves *Mercenaria*, *Mytilus* (Towe and Thompson, 1972) and *Pinctada*

(Nakahara and Bevelander, 1970) showed the presence of both a 'frothy' structure *within* single crystals and a 'bright layer' *between* crystalline units. This is a crucial observation which, supported by experimental data on the effect of a strong bleaching treatment on the organic matrix in shells (Crenshaw, 1972), allowed researchers to put forward the hypothesis that there were two pools of organics within biominerals: an intracrystalline matrix and an intercrystalline matrix. This concept was explored with regard to biomineralization and the organic–inorganic interface (Albeck et al., 1996) but went largely undetected in the field of geochemistry and especially protein diagenesis dating, as we will see in Chapter 4.

1.3 How and Why are Biominerals Formed?

Early work on biomineralized organisms, therefore, showed that building a mineralized skeleton involves a sophisticated balance between organic and inorganic matrices. In general, a high level of order can be found within the mineral skeleton, from nanometre to macroscopic length scales (Vielzeuf et al., 2008; Arakaki et al., 2015) although amorphous exceptions are known, e.g. in earthworm granules (Hodson et al., 2015). This extraordinary architecture is formed under physiological conditions, i.e. at temperatures and pressures that are lower than those required to produce the same materials by chemical synthesis (Wang and Nilsen-Hamilton, 2013). Since this implies an energetic cost, possessing a hard skeleton must be sufficiently advantageous as to warrant this investment (Lowenstam and Weiner, 1989; Knoll, 2003; Murdock and Donoghue, 2011). Indeed, the fact that this strategy pays off is shown by the fact that possessing a biomineralized skeleton or tissue is not a prerogative of a single lineage of organisms: corals, molluscs, turtles and humans are all able to build biogenic minerals (Table 1.1, adapted from Lowenstam, 1980; Figure 1, adapted from Murdock and Donoghue, 2011).

This peculiarity has captured the interest not only of biomineralization scientists, but also of evolutionary biologists, who are concerned with a fundamental question (Murdock and Donoghue, 2011): was there a common ancestor which developed the ability to mineralize a skeleton, or are we looking at independent evolution pathways for each lineage? The fact that biomineralizing groups are dispersed among other groups that do *not* build hard skeletons seems to suggest that the evolution of this ability was independent (Figure 1.1). Indeed, most biomineralized skeletons appear during the 'Cambrian explosion' as the result of the diversification of an original biomineralization 'episode' that had occurred during the (highly alkaline) Ediacaran (635–541 Ma ago). In the Ediacaran, most taxa were soft bodied, but some weakly calcified taxa also began to appear (*Cloudina*, *Namacalathus*, *Namapoikia*), supporting the idea of the diversification of an original feature, likely a 'biomineralization toolkit' (Livingston et al., 2006; Ramos-Silva et al., 2013; Wood et al., 2017).

But how can mineralization actually occur within biological systems? Early views based on classical nucleation theory posited that the early stages of

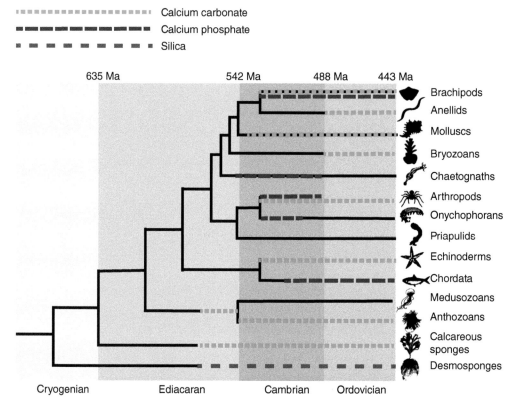

················ Calcium carbonate
━ ━ ━ ━ ━ ━ ━ ━ Calcium phosphate
▪ ▬ ▬ ▬ ▬ ▬ ▬ ▪ Silica

Figure 1.1 Evolutionary biology has had a longstanding interest in biomineralization. The phylogenetic tree shows that most biomineralized organisms appear in the Cambrian fossil record (i.e. the 'Cambrian explosion') and that mineralized skeletons are most frequently composed of calcium carbonate, calcium phosphate or silica. Adapted from Murdock and Donoghue (2011).

mineral formation are governed by random collisions between dissolved ions in supersaturated homogeneous solutions to form metastable clusters, which can then disintegrate further or grow (Gebauer et al., 2014). Recent mechanistic and experimental studies have shown that this is not possible; given the energetic cost involved in mineralization under physiological conditions, a biological catalyst is needed: the 'biomolecular toolkit'. This initiated the interaction between inorganic ions (calcium, carbonate, phosphate) and the extracellular matrix in the first biomineralized organisms ca. 550 Ma ago (Wood et al., 2017). It is likely that this toolkit used to be a relatively simple one but, after many million years of evolution, has morphed into an extremely complex 'molecular machinery', i.e. the organic matrix we find in biominerals today. This is so difficult to characterize fully that, as a result, many of the fundamental mechanisms controlling biogenic mineralization are still fairly obscure (De Yoreo et al., 2015). However, the exponential growth in the number of studies on biomineralization during the past three decades (Figure 1.2) has

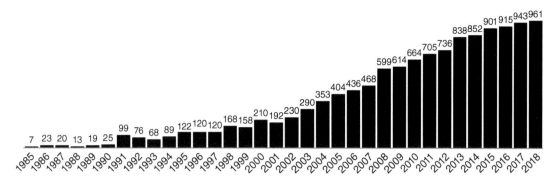

Figure 1.2 Number of articles with keyword 'biomin*' published between 1985 and 2018 (source: Web of Science).

so far allowed us to clarify three aspects of biologically controlled mineralization (Mann, 1983).

Biomineralization:

i is regulated by the genome;

ii results in the formation of mineral phases that are not in equilibrium with the environment;

iii is modulated by an organic matrix (extracellular) which is then entrapped in the mineral phase (Marin et al., 2012) and which can be recovered in biominerals of geological ages (Abelson, 1954).

In practice, this knowledge has changed our understanding of the way in which minerals are formed (Gebauer et al., 2008): new theories propose that, initially, ions assemble into stable amorphous clusters ('prenucleation clusters'), which can be conceived of as highly dynamic polymers of ion pairs. Subsequently, via a multi-step sequence, clusters are formed, followed by amorphous intermediates and, finally, crystals. Organic species, and especially proteins, can efficiently regulate many of these steps (Evans, 2013). In fact, the progress from ions (or atoms and molecules) to crystals can occur via a range of mechanisms, summarized recently by De Yoreo et al. (2015) (Figure 1.3).

In summary, biominerals are a diverse group of biological entities, which have a long evolutionary history, spanning at least 550 million years, and which display a fascinating set of coordinated biochemical mechanisms whereby macromolecules promote and regulate the growth of a mineralized skeleton. Proteins in particular play a major role in this process. This interplay not only confers exceptional material properties to the biogenic mineral – a feature which modern engineers strive to replicate in the creation of biomimetic and bioinspired materials – but also guarantees it a better chance of survival post mortem, in the fossil record.

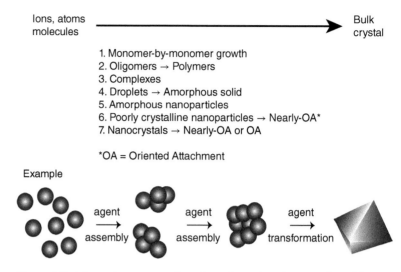

Ions, atoms molecules → Bulk crystal

1. Monomer-by-monomer growth
2. Oligomers → Polymers
3. Complexes
4. Droplets → Amorphous solid
5. Amorphous nanoparticles
6. Poorly crystalline nanoparticles → Nearly-OA*
7. Nanocrystals → Nearly-OA or OA

*OA = Oriented Attachment

Example

agent assembly   agent assembly   agent transformation

Figure 1.3 Schematic representation of nonclassical nucleation, adapted from Evans (2013) and De Yoreo et al. (2015).

1.4 'Biomineralization Toolkit': From Proteins to Proteomes

The fact that proteins are an important (but not the only) part of the biomineralization toolkit has been known for at least two decades (Sigel et al., 2008) and early biomolecular research efforts in the field were directed towards isolating, purifying and sequencing these molecules from a range of biominerals in order to clarify just how similar the 'biomineralization toolkit' is across various phyla. The first organisms and tissues to be investigated were mollusc shell (especially taxa producing pearls), avian eggshell and animal bone/tooth, due to their commercial and medical relevance. The study of these first sequences showed that, at least in mollusc shell, there were typically two subsets of organic matrices: the acid-soluble and the acid-insoluble (protein, proteoglycan and chitin) matrices (Miyamoto et al., 1996; Sarashina and Endo, 1998; Mann et al., 2000; Weiss et al., 2000; Marin et al., 2007; Marie et al., 2008). Furthermore, these early analyses showed that different proteins could be isolated from different microstructures and mineralogies (e.g. prisms and nacre/mother-of-pearl in mollusc shells) (Marie et al., 2012). Some similarities also began to emerge: C-lectin type proteins were isolated and purified from *Haliotis* shell (Weiss et al., 2000), avian eggshell (Mann and Siedler, 1999, 2004; Mann, 2004), sea urchins (Illies et al., 2002); highly acidic proteins were found commonly in mollusc shells (Gotliv et al., 2005; Marie et al., 2007) and a large percentage of them are intrinsically disordered proteins (IDPs), which are thought to be important for initiating biomineralization (Boskey and Villarreal-Ramirez, 2016).

Knowledge of molecular sequences is not sufficient for understanding the way in which these proteins (and their three-dimensional structures) interact with inorganic ions, clusters, amorphous precursors and crystals throughout the process of mineralization. Mathematical modelling can be used effectively to simulate the organic–inorganic interactions at each stage and is therefore able to elucidate the role of a certain biomolecule. An example is the study of the chicken eggshell C-lectin Ovocleidin-17 (OC-17), which showed that this molecule can bind to the carbonate ions of amorphous calcium carbonate via basic (positively charged) amino acid residues and is then released from the surface of the formed calcite crystal to start its catalytic cycle once more (Freeman et al., 2011). An important proviso of these types of computational studies is that not only the protein sequence, but also its structure, needs to be known.

An alternative strategy to computer simulation is the experimental approach, which often involves purifying a protein and then observing the behaviour of the mineral phase in vitro (Falini et al., 1996). It is also possible to create synthetic proteins (or peptides) with domains of known primary sequence and structural properties, e.g. rigid lectin-like domains versus intrinsically disordered. For example, a study of the three domains of the sea urchin spicule matrix protein SM50, using a small ubiquitin-like modifier (SUMO) fusion protein as a model, has shown that the three domains are necessary and sufficient to allow the formation and growth of the sea urchin spicule (Rao et al., 2013). In an earlier study, recombinant nacrein proteins with and without carbonic anhydrase and Gly–X–Asn domains were tested in order to elucidate nacrein's role as a negative regulator of calcification (Miyamoto et al., 2005).

Recent efforts in molecular biology and the sequencing of hundreds of new organisms, including biomineralizing taxa, have shown that the number and variety of proteins trapped in even apparently simple systems (e.g. avian eggshell) is such that a single-molecule approach is no longer viable. As such, it would be more correct to refer to the proteins pertaining to biominerals as 'biomineralized proteomes'. Chapter 3 briefly summarizes the main protein families found in biominerals. It is worth mentioning here, though, a recent review of the 'secretome' of marine mollusc shells by Kocot et al. (2016), which highlights the extraordinary diversity of biomolecules secreted by the molluscs' mantle cells in order to modulate the calcification of the mineral matrix, including mechanical and aesthetic characteristics (e.g. hardness, colour). Interestingly, it appears that most proteins involved in biomineralization (not only in shells but also in other animals) display repetition of low complexity domains (RLCDs), which are easily shuffled, lost or acquired in evolutionary processes (Kocot et al., 2016).

The quest for the universal biomineralization toolkit is at a crucial stage, with more and more proteins being identified and their function elucidated by homology. It is clear that there is an interplay between different proteins and/or different protein domains, and that the study of single molecules can only give a partial picture of this complex molecular machinery. It is also clear that only some of the proteins belonging to the proteome perform a mineralization-related task. There is need, however, for a novel strategy able to select protein sequences that are truly relevant for biomineralization and distinguish them from those which only 'happen' to be incorporated in biominerals. Such an approach might stem from the study of fossil proteomes, i.e. the surviving proteins within (sub)fossil biominerals.

Box 1.1 Terminology

Amino acids are aminoalkanoic acids, each containing a hydrogen atom –H, an amino group –NH$_2$, a carboxylic group –COOH and an –R group, or side-chain, bound to a central carbon atom –C. The amine and carboxylic acid functional groups mean that amino acids can act both as an acid and as a base. The –R group is specific to each of the 20 common, standard or coded amino acids. Amino acids are usually referred to by trivial names (e.g. alanine), but a three-letter abbreviation system is commonly used (e.g. Ala) for amino acids which make up peptides and proteins, as well as a one-letter system (e.g. A) for identifying peptides through a string of letters.

ELECTRICALLY CHARGED SIDE CHAIN

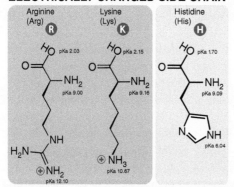

POLAR UNCHARGED SIDE CHAIN

SPECIAL CASES

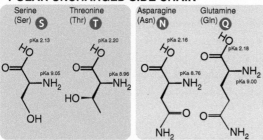

HYDROPHOBIC SIDE CHAIN

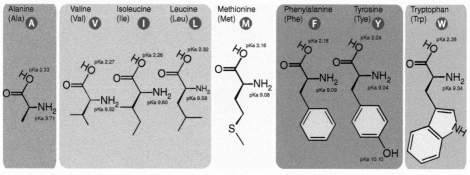

Image modified from the original at https://commons.wikimedia.org/wiki/File:Amino_Acids.svg and shared under a Creative Commons Attribution-Share Alike 3.0 Unported licence.

Peptides are formed by condensation of the respective amino and carboxy groups of two different amino acids, with the elimination of a water molecule and the formation of an amino linkage (–CO–NH–) (also known as a peptide bond). Generally, peptides are considered to be between two and twenty amino acids long (Barrett and Elmore, 1998). Proteins are polymers of amino acids with molecular weights in the 1–1000 kDa region, although the size boundary between peptide and protein is not absolute: there are some very small proteins (osteocalcin, 100 residues long; insulin, 51 residues) and also some very long peptides (amyloid beta, 36–43 residues). Proteins and peptides generally retain an unreacted amino group at one end (N terminus) and an unreacted carboxyl group at the other (C terminus) and, at neutral pH, each terminus carries an ionic charge. Conventionally, amino acid residues in peptide chains are numbered from the N terminus to the C terminus. They are also written from left to right, since this is the direction followed during biosynthesis.

The number, type and sequence of amino acids in the chain determines the *primary* structure of the protein, while the three-dimensional configuration of the peptide chains in space is called the *secondary* structure (e.g. alpha-helices, beta-strands and beta-sheets). Further interactions and coiling through ionic interactions, hydrogen bonds, van der Waals forces or sulfur bridges can bring distant portions of the primary and secondary structures close together, thus determining the formation of a *tertiary* structure. The *quaternary* structure involves the association of two or more polypeptide chains into a multi-subunit, or oligomeric, protein (Barrett and Elmore, 1998). Crucial for the diversification of the protein's functions are post-translational modifications (PTMs): functional groups can be added (as in phosphorylation, methylation, glycosylation, acetylation) or parts of the protein can be degraded (for example enzymatically), in order to achieve the normal functioning of the living cell.

The source of information on nucleotide and protein sequences is the NCBI (National Center for Biotechnology Information: https://www.ncbi.nlm.nih.gov/) repository, which is also linked to the protein databank (PDB: http://www.rcsb.org/pdb), in which any protein structure that has been solved is available. It is important to note that proteins enclosed in biominerals are often not well characterized, mainly due to missing or incomplete genomic information for the biomineralizing organism. In the example below, the cDNA sequence of OC-17, a major biomineralization protein (Lakshminarayanan et al., 2002; Freeman et al., 2011) which had been sequenced directly in 1999 and the structure of which has been known since 2004 (Mann and Siedler, 1999; Reyes-Grajeda et al., 2004), was missing from the chicken genome and had to be determined using transcriptome assembly and cloning in 2014 (Zhang et al., 2014).

cDNA sequence of Ovocleidin-17, a major biomineralization protein from chicken eggshell (Zhang et al., 2014)

```
  1 GGTGCCGGGA CGCGGAACGG CCGCCCGAAC GGGAACGCAA TGGCACCGAC GTGGGCGCTG
 61 CTGGGCTGCG TTCTGCTGCT CCCCTCCCTG CGAGGGGATC CGGACGGCTG CGGCCCGGGT
121 TGGGTGCCGA CCCCCGGCGG CTGCCTCGGC TTCTTCAGCC GGGAGCTCAG CTGGAGCCGC
181 GCCGAGTCGT TCTGCCGCCG TTGGGGTCCC GGTTCCCACC TGGCGGCGGT GCGCAGCGCG
241 GCGGAGCTGC GGCTCCTCGC GGAGCTCCTC AACGCGTCGC GGGGCGGCGA CGGCAGCGGG
301 GAGGGGGCGG ACGGCCGCGT CTGGATCGGC CTCCACCGCC CCGCCGGGAG CCGTTCGTGG
361 CGGTGGTCGG ACGGCACCGC GCCGCGCTTC GCTTCGTGGC ACCGAACGGC CAAAGCGCGG
421 CGGGGGGGGC GGTGCGCGGC GCTGCGGGAC GAGGAGGCCT TCACCTCGTG GGCCGCCCGG
481 CCGTGCACAG AGCGCAATGC CTTCGTCTGC AAAGCCGCCG CCTGAATGGA CAACAACACA
541 ACAACACAAC AACACAACAA CGCAACAACG CAACAACGCA ACAACGCAAC GACCCCCAAC
601 ACTGCAATAA ACGGACCCAC AGCAGC
```

Primary (amino acid) sequence of OC-17 (Mann and Siedler, 1999)

```
  1 MAPTWALLGC VLLLPSLRGD PDGCGPGWVP TPGGCLGFFS RELSWSRAES FCRRWGPGSH
 61 LAAVRSAAEL RLLAELLNAS RGGDGSGEGA DGRVWIGLHR PAGSRSWRWS DGTAPRFASW
121 HRTAKARRGG RCAALRDEEA FTSWAARPCT ERNAFVCKAA A
```

Secondary (alpha-helix, beta-sheets and random coil configurations represented in red, yellow and light grey, respectively) and tertiary structure of OC-17 (Reyes-Grajeda et al., 2004).

1.5 Fossil Biominerals, Fossil Proteomes

Biomineralized organisms have a greater chance of entering and being preserved in the fossil record than do soft tissues, to the point that the typical 'fossil' is almost shorthand for 'fossilized skeletal tissue' of a variety of organisms (the ones classified in Table 1.1). The importance of these fossils for archaeology and earth sciences is obvious: they allow us to reconstruct the past, from deep geological times (let us just think of the 'index' or 'guide' fossils used to define and identify a geological period) until more recently (when clade *Homo* enters the world scene in the Quaternary, see Chapter 4).

The survival of organic matter in 'fossil' organisms is a discovery that dates back to the 19th century (Schaffer, 1889). Identification of these compounds as individual amino acids (Abelson, 1954), and of their chiral properties, led to the consequent development of a relative dating technique (amino acid racemization, AAR). The field of amino acid biogeochemistry grew very rapidly, reaching a pinnacle with regard to the integration of geochronological applications, biogeochemistry and biomineralization at a symposium held in 1978 in Virginia. The papers are collected in the text 'Biogeochemistry of amino acids' (Hare et al., 1980), which (remarkably) is still today an exceptional source of knowledge (or indeed of open questions) for palaeobiogeochemists.

Subsequent advances focused specifically on the potential use of different biominerals as substrates for protein diagenesis and geochemistry. This potential derives not only from the state of preservation, but also from the specific characteristics of the biomineral. In particular, biominerals can be classified into three broad categories with regard to their (generalized) behaviour towards protein diagenesis: open-system biominerals, closed-system biominerals and biominerals which may contain a closed system.

1.5.1 Open and Closed Systems: Isolating the Intracrystalline Organic Fraction in the Laboratory

Most biominerals display a complex microstructure (crystals take the form of prisms, platelets, cones, lamellae or needles) within which the organic framework is interspersed in a variety of ways. Despite this array of microstructural arrangements, the organic component can be classified more simply in terms of its spatial relationship with the crystals, whatever shape these might be. In the 1960s and 1970s amino acid geochronologists (including Hare, Wehmiller, King and Ritter) discussed the idea that the organic framework had components of different mobility (Wehmiller, 1980). Sykes et al. (1995) put forward a model that theorized the existence of two pools of proteins: one 'intercrystalline', which is located *between* mineral crystals and shows an open-system behaviour (i.e. leaching of amino acids out of the system, poor reproducibility of the data, contamination issues); the second 'intracrystalline', which is somewhat trapped *within* the crystals, can be isolated via a chemical treatment (strong oxidation, usually by bleaching with sodium hypochlorite, NaOCl) and can behave as a closed system (Collins and Riley, 2000; Penkman et al., 2008; Bright and Kaufman, 2011; Demarchi et al., 2013). Figure 1.4 shows this theoretical model in which an 'operational intracrystalline fraction' can be isolated.

In reality, of course, things are more complex, but recent visualization studies performed using a variety of techniques based on electron microscopy and tomography (Robach et al., 2005; Gries et al., 2009; Gordon and Joester, 2011; Li et al., 2011; Younis et al., 2012; Hendley et al., 2015; Liu et al., 2015, 2017) show that Towe and Thompson's early work (Towe and Thompson, 1972) was accurate and that the model by Sykes and colleagues stands. The main characteristics of the organic intracrystalline matrix thus revealed are:

- it is embedded within *single crystals*, i.e. biogenic crystals fracture and diffract as single crystals (see a review of evidence in Weber and Pokroy, 2015);
- the occlusions have different shapes (voids, islets, chain-like, haze-like) at the nanometre scale (for a summary see Liu et al., 2017);
- the protein component of the intracrystalline matrix tends to be highly acidic and often glycosylated (Marin et al., 2007);
- the conformation of these proteins tends to be open, unfolded, with 'intrinsically disordered domains' (Boskey and Villarreal-Ramirez, 2016).

In fossil biominerals the intercrystalline organic matrix, which is not contained within occlusions, will be the first to succumb to the effects of time and of the burial environment, and even if its degradation is somewhat halted, this fraction is likely to contain exogenous compounds: if a biomineral contains mainly intercrystalline proteins, it will display an open-system behaviour. Open-system biominerals are prone to contamination and leaching of amino acids from the system (Brooks et al., 1990; Sykes et al., 1995), and the proteins

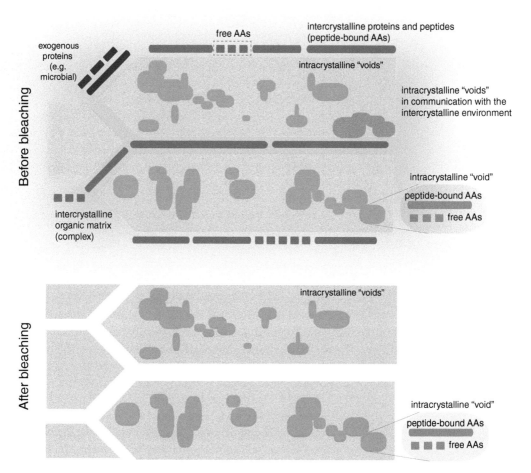

Figure 1.4 The intracrystalline fraction of proteins (peptide-bound and free amino acids) is occluded in nanometric 'voids' within the crystal. The morphology of the occlusions is variable: voids, islets, chain-like, haze-like shapes (Liu et al., 2015). The intercrystalline organic matrix also contains peptide-bound and free amino acids (as well as other compounds formed with the nonproteinaceous components of the matrix, not shown in the figure, and contamination from exogenous proteins) but these can be removed by a strong oxidation pretreatment (e.g. bleaching). This treatment leaves behind only the original proteins, degraded in situ. Modified after Sykes et al. (1995).

contained in an open system are not protected by the action of environmental factors such as pH, presence of water, effect of clay surfaces, of buffers, of metal ions (Smith and Evans, 1980). One example of an open-system biomineral is bone, since the individual mineral elements are too small to encapsulate proteins. In the case of bone, it may be that proteins become trapped in mineral aggregates (Salamon et al., 2005), though this has not yet been the focus of detailed studies. The fact that the organic matrix within bones behaves as an open system can be a real hindrance to archaeological and palaeontological applications since, without in-depth analyses, it is extremely difficult to assess whether we are looking at the original indigenous molecules, or at modern (or ancient) contamination. This is especially problematic when bulk amino acids are extracted from the bone matrix, typically for measuring the extent of racemization for geochronological purposes (e.g. Bada, 1985; see also Chapter 4). Later applications of ancient DNA and palaeoproteomics have been successful in extracting biomolecular *sequences*, which can be authenticated and distinguished from contamination (e.g. Ostrom et al., 2000; Höss et al., 1996), thus reducing the impact of the open-system behaviour of bone.

Conversely, early work on avian eggshells (especially ratite) postulated that they retain both original organic matter and degradation products during diagenesis (Brooks et al., 1990), and avian eggshell has been classed as a closed system. However, later research demonstrated that even avian eggshell is more akin to systems in which a *fraction* of proteins behaving as a closed system can be isolated from the complex organic matrix (Crisp et al., 2013). It is likely that the extremely rapid formation of the eggshell in utero (~20 hours for chicken eggshell, see Figure 3.3) means that a large proportion of proteins is trapped within the large eggshell crystals; this happens fairly indiscriminately, as indicated by the complexity of the eggshell proteome (Mann, 2015; Demarchi et al., 2016). Eggshell has been the traditional target for amino acid racemization dating from the 1990s onwards, and more recent studies in palaeoproteomics have demonstrated that eggshell offers an ideal substrate for exceptional preservation (Demarchi et al., 2016), but also for interesting archaeological insights into the relationship between humans and their environment (Stewart et al., 2014; Jonuks et al., 2018; Demarchi et al., 2019).

Systematic investigation of a range of biominerals has shown that a functionally defined intracrystalline *fraction* can be isolated from many taxa of mollusc shells, corals, brachiopods and ostracods (Walton, 1998; Penkman et al., 2008, 2011; Bright and Kaufman, 2011; Hendy et al., 2012; Demarchi et al., 2013; Bosch et al., 2015; Pierini et al., 2016). Unlike eggshell, these tissues and organisms are formed over longer time spans, even tens of hundreds of years: for example, the oldest animal in the world was an *Arctica islandica* bivalve mollusc, which was 'fished' from the seafloor when it was 507 years old, thus making it the world's oldest animal (Butler et al., 2013). These organisms, therefore, are much more complex, because they will include 'old' proteins due to lack of tissue turnover, as well as 'young'.

The isolation of the intracrystalline fraction in the laboratory, despite taking many decades to become standard procedure, is in practice a very simple step (see Chapter 4): the biomineral is cleaned (whole or in fragments, depending on its size) by ultrasonication in water or by briefly soaking in a weak acid (e.g. acetic acid), base (e.g. sodium hydroxide), or calcium-chelating agent (ethylenediaminetetraacetic acid). For shells, it is common to separate different microstructural layers by abrasion (Marie et al., 2012), after which the selected mineral fraction is ground and sieved until a homogeneous particle size (usually <500 μm) is obtained. The dried powders are submerged in an excess of oxidizing agent (e.g. NaOCl, 12% wt/vol), often helping the bleaching process by gently agitating the suspension.

Powders are rinsed in water and methanol (e.g. Penkman et al., 2008) until all the bleach has been removed. The bleached powders should theoretically only contain the intracrystalline proteins embedded in the mineral. The next step is therefore to extract and characterize these proteins and use them for a variety of applications in archaeology and earth sciences. However, the results vary slightly according to the experimental settings used (e.g. the concentration of the oxidant, the duration of the treatment, and the optimal particle size need to be tested for each biominerals) and this indicates that the intracrystalline fraction is indeed not a perfect closed system (nor is always the same in all biological systems, even for two shells of the same species). Therefore, the operational definition usually refers to crushed biomineral (around 500 μm particle size) exposed to 12% wt/vol NaOCl for 48 hours (Penkman et al., 2008).

1.5.2 'Diagenetically isolated' intracrystalline proteins

The concept of intracrystalline proteins has proven to be extremely powerful both for biomineralization studies (Drake et al., 2013; Ramos-Silva et al., 2013) and for geochronology/palaeoproteomics (Chapters 4 and 5). In fossils, however, the situation is complicated due to the matter of degradation: entombed proteins cannot be expected to survive the effects of time untouched (even permafrost-buried remains degrade, albeit slowly). This complication is, at the same time, a positive factor: diagenesis acts as an effective agent for isolating the intracrystalline fraction, by removing the more 'mobile' components and leaving behind a stable system (see e.g. Wehmiller and Miller, 2000). Therefore, the many studies that used geological and archaeological samples for dating purposes from the 1960s onwards without a bleaching step were successful *because the analyses effectively targeted a closed system* (Chapter 4). Re-evaluating those early datasets, preferably those obtained from well-understood stratigraphic sequences, by analysing bleached and unbleached samples, would be an excellent way to gain a better understanding of the behaviour of different fractions of proteins.

As a final remark to this chapter, it is important to highlight that, in general, the information we retrieve from fossil biominerals will be biased towards the most degradation-resistant molecules. Therefore, it is crucial that the mechanisms of diagenesis and preservation are understood before embarking on a discussion of the potential applications.

References

Abelson, P.H. (1954). Amino acids in fossils. *Science* 119: 576.

Abelson, P.H. (1956). Paleobiochemistry. *Sci. Am.* 195: 83–96.

Addadi, L. and Weiner, S. (2014). Biomineralization: Mineral formation by organisms. *Phys. Scr.* 89: 098003.

Albeck, S., Addadi, I., and Weiner, S. (1996). Regulation of calcite crystal morphology by intracrystalline acidic proteins and glycoproteins. *Connect. Tissue Res.* 35: 365–370.

Arakaki, A., Shimizu, K., Oda, M. et al. (2015). Biomineralization-inspired synthesis of functional organic/inorganic hybrid materials: Organic molecular control of self-organization of hybrids. *Org. Biomol. Chem.* 13: 974–989.

Asara, J.M., Schweitzer, M.H., Freimark, L.M. et al. (2007). Protein sequences from mastodon and *Tyrannosaurus rex* revealed by mass spectrometry. *Science* 316: 280–285.

Bada, J.L. (1985). Aspartic acid racemization ages of California Paleoindian Skeletons. *Am. Antiq.* 50: 645–647.

Barrett, G.C. and Elmore, D.T. (1998). *Amino Acids and Peptides.* Cambridge University Press.

Bosch, M.D., Mannino, M.A., Prendergast, A.L. et al. (2015). New chronology for Ksâr 'Akil (Lebanon) supports Levantine route of modern human dispersal into Europe. *Proc. Natl. Acad. Sci. U.S.A.* 112: 7683–7688.

Boskey, A.L. and Villarreal-Ramirez, E. (2016). Intrinsically disordered proteins and biomineralization. *Matrix Biol.* 52–54: 43–59.

Bright, J. and Kaufman, D.S. (2011). Amino acid racemization in lacustrine ostracodes, part I: Effect of oxidizing pre-treatments on amino acid composition. *Quat. Geochronol.* 6: 154–173.

Brooks, A.S., Hare, P.E., Kokis, J.E. et al. (1990). Dating Pleistocene archeological sites by protein diagenesis in ostrich eggshell. *Science* 248: 60–64.

Buckley, M., Walker, A., Ho, S.Y.W. et al. (2008). Comment on 'Protein sequences from mastodon and *Tyrannosaurus rex* revealed by mass spectrometry.' *Science* 319: 33.

Buckley, M., Warwood, S., van Dongen, B. et al. (2017). A fossil protein chimera; difficulties in discriminating dinosaur peptide sequences from modern cross-contamination. *Proc. R. Soc. B* 284: 20170544.

Butler, P.G., Wanamaker, A.D., Scourse, J.D. et al. (2013). Variability of marine climate on the North Icelandic Shelf in a 1357-year proxy archive based on growth increments in the bivalve *Arctica islandica*. *Palaeogeogr. Palaeoclimatol. Palaeoecol.* 373: 141–151.

Cappellini, E., Prohaska, A., Racimo, F. et al. (2018). Ancient biomolecules and evolutionary inference. *Annu. Rev. Biochem.* 87: 1029–1060.

Cappellini, E., Welker, F., Pandolfi, L., and Madrigal, J.R. (2019). Early Pleistocene enamel proteome from Dmanisi resolves Stephanorhinus phylogeny. *Nature* 574: 103–107.

Carter, J.G. (ed.) (1989). *Skeletal Biomineralization: Patterns, Processes and Evolutionary Trends*, vol. 5. American Geophysical Union.

Cleland, T.P., Schroeter, E.R., Zamdborg, L. et al. (2015). Mass spectrometry and antibody-based characterization of blood vessels from *Brachylophosaurus canadensis*. *J. Proteome Res.* 14: 5252–5262.

Collins, M.J. and Riley, M.S. (2000). Amino acid racemization in biominerals: The impact of protein degradation and loss. In: *Perspectives in Amino Acid and Protein Geochemistry* (eds. G. Goodfriend, M.J. Collins, M. Fogel, et al.), 120–141. Oxford University Press.

Collins, M.J., Gernaey, A.M., Nielsen-Marsh, C.M. et al. (2000). Slow rates of degradation of osteocalcin: Green light for fossil bone protein? *Geology* 28: 1139–1142.

Crenshaw, M.A. (1972). The soluble matrix from *Mercenaria mercenaria* shell. *Biomineralization* 6: 6–11.

Crisp, M., Demarchi, B., Collins, M. et al. (2013). Isolation of the intra-crystalline proteins and kinetic studies in *Struthio camelus* (ostrich) eggshell for amino acid geochronology. *Quat. Geochronol.* 16: 110–128.

Demarchi, B., Rogers, K., Fa, D.A. et al. (2013). Intra-crystalline protein diagenesis (IcPD) in *Patella vulgata*. Part I: Isolation and testing of the closed system. *Quat. Geochronol.* 16: 144–157.

Demarchi, B., Hall, S., Roncal-Herrero, T. et al. (2016). Protein sequences bound to mineral surfaces persist into deep time. *eLife*, 5: e17092.

Demarchi, B., Presslee, S., Gutiérrez-Zugasti, I. et al. (2019). Birds of prey and humans in prehistoric Europe: A view from El Mirón Cave, Cantabria (Spain). *J. Archaeol. Sci. Rep.* 24: 244–252.

De Yoreo, J.J., Gilbert, P.U.P.A., Sommerdijk, N.A.J.M. et al. (2015). Crystal growth. Crystallization by particle attachment in synthetic, biogenic, and geologic environments. *Science* 349: aaa6760.

Drake, J.L., Mass, T., Haramaty, L. et al. (2013). Proteomic analysis of skeletal organic matrix from the stony coral *Stylophora pistillata*. *Proc. Natl. Acad. Sci. U.S.A.* 110: 3788–3793.

Emiliani, C. (1955). Mineralogical and chemical composition of the tests of certain Pelagic Foraminifera. *Micropaleontology* 1: 377–380.

Evans, J.S. (2013). 'Liquid-like' biomineralization protein assemblies: A key to the regulation of non-classical nucleation. *CrystEngComm* 15: 8388.

Falini, G., Albeck, S., Weiner, S. et al. (1996). Control of aragonite or calcite polymorphism by mollusk shell macromolecules. *Science* 271: 67–69.

Farre, B. and Dauphin, Y. (2009). Lipids from the nacreous and prismatic layers of two Pteriomorpha Mollusc shells. *In: EGU General Assembly Conference Abstracts* 11: 5447.

Freeman, C.L., Harding, J.H., Quigley, D., and Rodger, P.M. (2011). Simulations of ovocleidin-17 binding to calcite surfaces and its implications for eggshell formation. *J. Phys. Chem. C* 115: 8175–8183.

Gaffey, S.J. (1988). Water in skeletal carbonates. *J. Sediment. Res.* 58: 397–414.

Gebauer, D., Völkel, A., and Cölfen, H. (2008). Stable prenucleation calcium carbonate clusters. *Science* 322: 1819–1822.

Gebauer, D., Kellermeier, M., Gale, J.D. et al. (2014). Pre-nucleation clusters as solute precursors in crystallisation. *Chem. Soc. Rev.* 43: 2348–2371.

Gordon, L.M. and Joester, D. (2011). Nanoscale chemical tomography of buried organic-inorganic interfaces in the chiton tooth. *Nature* 469: 194–197.

Gotliv, B.-A., Kessler, N., Sumerel, J.L. et al. (2005). Asprich: A novel aspartic acid-rich protein family from the prismatic shell matrix of the bivalve *Atrina rigida*. *Chembiochem* 6: 304–314.

Goulletquer, P. and Wolowicz, M. (1989). The shell of *Cardium edule, Cardium glaucum* and *Ruditapes philippinarum*: Organic content, composition and energy value, as determined by different methods. *J. Mar. Biol. Assoc. U.K.* 69: 563–572.

Gries, K., Kröger, R., Kübel, C. et al. (2009). Investigations of voids in the aragonite platelets of nacre. *Acta Biomater.* 5: 3038–3044.

Hare, P.E., Hoering, T.C., and King, K. (eds.) (1980). *Biogeochemistry of Amino Acids*. New York: Wiley.

Hendley, C.T., Tao, J., Kunitake, J.A.M.R. et al. (2015). Microscopy techniques for investigating the control of organic constituents on biomineralization. *MRS Bulletin* 40: 480–489.

Hendy, E.J., Tomiak, P.J., Collins, M.J. et al. (2012). Assessing amino acid racemization variability in coral intra-crystalline protein for geochronological applications. *Geochim. Cosmochim. Acta* 86: 338–353.

Hendy, J., Welker, F., Demarchi, B. et al. (2018). A guide to ancient protein studies. *Nat. Ecol. Evol.* 2: 791–799.

Hodson, M.E., Benning, L.G., Demarchi, B. et al. (2015). Biomineralisation by earthworms: An investigation into the stability and distribution of amorphous calcium carbonate. *Geochem. Trans.* 16: 4.

Höss, M., Jaruga, P., Zastawny, T.H. et al. (1996). DNA damage and DNA sequence retrieval from ancient tissues. *Nucleic Acids Res.* 24: 1304–1307.

Hudson, J.D. (1967). The elemental composition of the organic fraction, and the water content, of some recent and fossil mollusc shells. *Geochim. Cosmochim. Acta* 31: 2361–2378.

Illies, M.R., Peeler, M.T., Dechtiaruk, A.M., and Ettensohn, C.A. (2002). Identification and developmental expression of new biomineralization proteins in the sea urchin *Strongylocentrotus purpuratus*. *Dev. Genes Evol.* 212: 419–431.

Jonuks, T., Oras, E., Best, J., and Demarchi, B. (2018). Multi-method analysis of avian eggs as grave goods: Revealing symbolism in conversion period burials at Kukruse, NE Estonia. *Environ. Archaeol.* 23: 109–122.

Knoll, A.H. (2003). Biomineralization and evolutionary history. *Rev. Mineral. Geochem.* 54: 329–356.

Kocot, K.M., Aguilera, F., McDougall, C. et al. (2016). Sea shell diversity and rapidly evolving secretomes: Insights into the evolution of biomineralization. *Front. Zool.* 13: 23.

Lakshminarayanan, R., Kini, R.M., and Valiyaveettil, S. (2002). Investigation of the role of ansocalcin in the biomineralization in goose eggshell matrix. *Proc. Natl. Acad. Sci. U.S.A.* 99: 5155–5159.

Lécuyer, C. and O'Neil, J.R. (1994). Stable isotope compositions of fluid inclusions in biogenic carbonates. *Geochim. Cosmochim. Acta* 58: 353–363.

Li, H., Xin, H.L., Kunitake, M.E. et al. (2011). Calcite prisms from mollusk shells (*Atrina rigida*): Swiss-cheese-like organic–inorganic single-crystal composites. *Adv. Funct. Mater.* 21: 2028–2034.

Liu, C., Xie, L., and Zhang, R. (2015). Heterogeneous distribution of dye-labelled biomineralizaiton proteins in calcite crystals. *Sci. Rep.* 5: 18338.

Liu, C., Du, J., Xie, L., and Zhang, R. (2017). Direct observation of nacre proteins in the whole calcite by super-resolution microscopy reveals diverse occlusion patterns. *Cryst. Growth Des.* 17: 1966–1976.

Livingston, B.T., Killian, C.E., Wilt, F. et al. (2006). A genome-wide analysis of biomineralization-related proteins in the sea urchin *Strongylocentrotus purpuratus*. *Dev. Biol.* 300: 335–348.

Lowenstam, H. (1980). Bioinorganic constituents of hard parts. In: *Biogeochemistry of Amino Acids* (eds. P.E. Hare, T.C. Hoering and K. King Jr.), 3–16. New York: Wiley.

Lowenstam, H.A. (1981). Minerals formed by organisms. *Science* 211: 1126–1131.

Lowenstam, H.A. and Weiner, S. (1989). *On Biomineralization*. Oxford University Press.

Luquet, G., Fernández, M.S., Badou, A. et al. (2012). Comparative ultrastructure and carbohydrate composition of gastroliths from Astacidae, Cambaridae and Parastacidae freshwater crayfish (Crustacea, Decapoda). *Biomolecules* 3: 18–38.

Mann, K. (2004). Identification of the major proteins of the organic matrix of emu (*Dromaius novaehollandiae*) and rhea (*Rhea americana*) eggshell calcified layer. *Br. Poult. Sci.* 45: 483–490.

Mann, K. (2015). The calcified eggshell matrix proteome of a songbird, the zebra finch (*Taeniopygia guttata*). *Proteome Sci.* 13: 29.

Mann, K. and Siedler, F. (1999). The amino acid sequence of ovocleidin 17, a major protein of the avian eggshell calcified layer. *Biochem. Mol. Biol. Int.* 47: 997–1007.

Mann, K. and Siedler, F. (2004). Ostrich (*Struthio camelus*) eggshell matrix contains two different C-type lectin-like proteins. Isolation, amino acid sequence, and posttranslational modifications. *Biochim. Biophys. Acta* 1696: 41–50.

Mann, K., Weiss, I.M., André, S. et al. (2000). The amino-acid sequence of the abalone (*Haliotis laevigata*) nacre protein perlucin. *FEBS J.* 267: 5257–5264.

Mann, S. (1983). Mineralization in biological systems. In: *Inorganic Elements in Biochemistry*, 125–174. Springer.

Marie, B., Luquet, G., Pais De Barros, J.-P. et al. (2007). The shell matrix of the freshwater mussel *Unio pictorum* (Paleoheterodonta, Unionoida). Involvement of acidic polysaccharides from glycoproteins in nacre mineralization. *FEBS J.* 274: 2933–2945.

Marie, B., Luquet, G., Bédouet, L. et al. (2008). Nacre calcification in the freshwater mussel *Unio pictorum*: Carbonic anhydrase activity and purification of a 95 kDa calcium-binding glycoprotein. *Chembiochem* 9: 2515–2523.

Marie, B., Joubert, C., Tayalé, A. et al. (2012). Different secretory repertoires control the biomineralization processes of prism and nacre deposition of the pearl oyster shell. *Proc. Natl. Acad. Sci. U.S.A.* 109: 20986–20991.

Marin, F., Luquet, G., Marie, B., and Medakovic, D. (2007). Molluscan shell proteins: Primary structure, origin, and evolution. *Curr. Top. Dev. Biol.* 80: 209–276.

Marin, F., Le Roy, N., and Marie, B. (2012). The formation and mineralization of mollusk shell. *Front. Biosci.* 4: 1099–1125.

Miller, M.F., 2nd and Wyckoff, R.W. (1968). Proteins in dinosaur bones. *Proc. Natl. Acad. Sci. U.S.A.*, 60: 176–178.

Miyamoto, H., Miyashita, T., Okushima, M. et al. (1996). A carbonic anhydrase from the nacreous layer in oyster pearls. *Proc. Natl. Acad. Sci. U.S.A.* 93: 9657–9660.

Miyamoto, H., Miyoshi, F., and Kohno, J. (2005). The carbonic anhydrase domain protein nacrein is expressed in the epithelial cells of the mantle and acts as a negative regulator in calcification in the mollusc *Pinctada fucata*. *Zoolog. Sci.* 22: 311–315.

Murdock, D.J.E. and Donoghue, P.C.J. (2011). Evolutionary origins of animal skeletal biomineralization. *Cells Tissues Organs* 194: 98–102.

Nakahara, H. and Bevelander, G. (1970). An electron microscope study of the muscle attachment in the mollusc *Pinctada radiata*. *Tex. Rep. Biol. Med.* 28: 279–286.

Ostrom, P.H., Schall, M., Gandhi, H. et al. (2000). New strategies for characterizing ancient proteins using matrix-assisted laser desorption ionization mass spectrometry. *Geochim. Cosmochim. Acta* 64: 1043–1050.

Penkman, K.E.H., Kaufman, D.S., Maddy, D., and Collins, M.J. (2008). Closed-system behaviour of the intra-crystalline fraction of amino acids in mollusc shells. *Quat. Geochronol.* 3: 2–25.

Penkman, K.E.H., Preece, R.C., Bridgland, D.R. et al. (2011). A chronological framework for the British Quaternary based on *Bithynia opercula*. *Nature* 476: 446–449.

Pierini, F., Demarchi, B., Turner, J., and Penkman, K. (2016). Pecten as a new substrate for IcPD dating: The Quaternary raised beaches in the Gulf of Corinth, Greece. *Quat. Geochronol.* 31: 40–52.

Ramos-Silva, P., Marin, F., Kaandorp, J., and Marie, B. (2013). Biomineralization toolkit: The importance of sample cleaning prior to the characterization of biomineral proteomes. *Proc. Natl. Acad. Sci. U.S.A.* 110: E2144–E2146.

Rao, A., Seto, J., Berg, J.K. et al. (2013). Roles of larval sea urchin spicule SM50 domains in organic matrix self-assembly and calcium carbonate mineralization. *J. Struct. Biol.* 183: 205–215.

Rao, A., Berg, J.K., Kellermeier, M., and Gebauer, D. (2014). Sweet on biomineralization: Effects of carbohydrates on the early stages of calcium carbonate crystallization. *Eur. J. Mineral.* 26: 537–552.

Reyes-Grajeda, J.P., Moreno, A., and Romero, A. (2004). Crystal structure of ovocleidin-17, a major protein of the calcified *Gallus gallus* eggshell: Implications in the calcite mineral growth pattern. *J. Biol. Chem.* 279: 40876–40881.

Robach, J.S., Stock, S.R., and Veis, A. (2005). Transmission electron microscopy characterization of macromolecular domain cavities and microstructure of single-crystal calcite tooth plates of the sea urchin *Lytechinus variegatus*. *J. Struct. Biol.* 151: 18–29.

Saitta, E.T., Liang, R., Lau, C.Y. et al. (2019). Cretaceous dinosaur bone contains recent organic material and provides an environment conducive to microbial communities. *eLife* 8: e46205.

Salamon, M., Tuross, N., Arensburg, B., and Weiner, S. (2005). Relatively well preserved DNA is present in the crystal aggregates of fossil bones. *Proc. Natl. Acad. Sci. U.S.A.* 102: 13783–13788.

Sarashina, I. and Endo, K. (1998). Primary structure of a soluble matrix protein of scallop shell: Implications for calcium carbonate biomineralization. *Am. Mineral.* 83: 1510–1515.

Schaffer, J. (1889). Über den feineren Bau fossiler Knochen. *Sitzungsberichte Kaiserlich Akad Wissenschaft Wien Math Naturwissenschaft Klasse* 3–7: 319–382.

Schroeter, E.R., DeHart, C.J., Cleland, T.P. et al. (2017). Expansion for the *Brachylophosaurus canadensis* collagen i sequence and additional evidence of the preservation of Cretaceous protein. *J. Proteome Res.* 16: 920–932.

Schweitzer, M.H., Marshall, M., Carron, K. et al. (1997). Heme compounds in dinosaur trabecular bone. *Proc. Natl. Acad. Sci. U.S.A.* 94: 6291–6296.

Sigel, A., Sigel, H., and Sigel, R.K.O. (2008). *Biomineralization: From Nature to Application*. John Wiley & Sons.

Smith, G.G. and Evans, R.C. (1980). The effect of structure and conditions on the rate of racemization of free and bound amino acids. In: *Biogeochemistry of Amino Acids* (eds. P.E. Hare, T.C. Hoering and K. King Jr.), 257–282. New York: Wiley.

Stern, B. (1996). Biomineral lipids in living and fossil molluscs. PhD thesis, Newcastle University.

Stewart, J., Allen, R.B., Jones, A.K.G. et al. (2014). Walking on eggshells: A study of egg use in Anglo-Scandinavian York based on eggshell identification using ZooMS. *Int. J. Osteoarchaeol.* 24: 247–255.

Sykes, G.A., Collins, M.J., and Walton, D.I. (1995). The significance of a geochemically isolated intracrystalline organic fraction within biominerals. *Org. Geochem.* 23: 1059–1065.

Towe, K.M. and Thompson, G.R. (1972). The structure of some bivalve shell carbonates prepared by ion-beam thinning. *Calcif. Tissue Res.* 10: 38–48.

Vielzeuf, D., Garrabou, J., Baronnet, A. et al. (2008). Nano to macroscale biomineral architecture of red coral (*Corallium rubrum*). *Am. Mineral.* 93: 1799–1815.

Walton, D. (1998). Degradation of intracrystalline proteins and amino acids in fossil brachiopods. *Org. Geochem.* 28: 389–410.

Wang, L. and Nilsen-Hamilton, M. (2013). Biomineralization proteins: From vertebrates to bacteria. *Front. Biol.* 8: 234–246.

Weber, E. and Pokroy, B. (2015). Intracrystalline inclusions within single crystalline hosts: From biomineralization to bio-inspired crystal growth. *CrystEngComm* 17: 5873–5883.

Wehmiller, J.F. (1980). Intergeneric differences in apparent racemization kinetics in mollusks and foraminifera: Implications for models of diagenetic racemization. In: *Biogeochemistry of Amino Acids* (eds. P.E. Hare, T.C. Hoering and K. King Jr.), 341–345. New York: Wiley.

Wehmiller, J.F. and Miller, G.H. (2000). Aminostratigraphic dating methods in quaternary geology. *Quat. Geochronol.*: 187–222.

Weiner, S. and Dove, P.M. (2003). An overview of biomineralization processes and the problem of the vital effect. *Rev. Mineral. Geochem.* 54: 1–29.

Weiss, I.M., Kaufmann, S., Mann, K., and Fritz, M. (2000). Purification and characterization of perlucin and perlustrin, two new proteins from the shell of the mollusc *Haliotis laevigata*. *Biochem. Biophys. Res. Commun.* 267: 17–21.

Welker, F. (2018). Palaeoproteomics for human evolution studies. *Quat. Sci. Rev.* 190: 137–147.

Wood, R., Ivantsov, A.Y., and Zhuravlev, A.Y. (2017). First macrobiota biomineralization was environmentally triggered. *Proc. Biol. Sci.*, 284: 20170059.

Younis, S., Kauffmann, Y., Bloch, L., and Zolotoyabko, E. (2012). Inhomogeneity of nacre lamellae on the nanometer length scale. *Cryst. Growth Des.* 12: 4574–4579.

Zhang, Q., Liu, L., Zhu, F. et al. (2014). Integrating de novo transcriptome assembly and cloning to obtain chicken Ovocleidin-17 full-length cDNA. *PLoS One* 9: e93452.

2

Mechanisms of Degradation and Survival

2.1 Introduction

'Diagenesis' is a geological term, which was first used within the field of carbonate sedimentology (and later palaeobiogeochemistry), to describe the complex network of physical processes and chemical reactions that occur post mortem in the burial environment and that ultimately transform a living organism into its constituent atoms. In the case of biominerals, there will be diagenetic processes driving the transformation of both the mineral and organic phases, and it is crucial to remember that the two cannot occur independently – the breakdown of the inorganic phase will influence the way in which the organic phase is degraded (or indeed preserved) and vice versa.

However, the complexity of the diagenetic network is such that, typically, the processes of organic degradation are considered as if they were occurring in isolation (in a 'liquid bubble') and the organic–inorganic interface tends to be ignored. Furthermore, the organic fraction of biominerals is composed of different classes of organic molecules (carbohydrates, lipids and proteins), each of which is usually considered independently. Within each of these classes, there is considerable variability, e.g. 'proteins' are composed of hundreds of widely different sequences and structures. (A summary of the main protein families in biominerals will be given in Chapter 3.) As a result, diagenesis is not a simple process. Hoering (1980) attempted to put forward a general scheme and proposed at least five pathways of diagenesis for organic matter in fossil mollusc shells (Scheme 2.1); despite 40 years having elapsed since this schematic view of diagenesis was posited, little experimental work has been carried out in order to verify its hypotheses.

Keeping this complexity in mind is important, especially when we attempt to describe the patterns of diagenesis mathematically. But why should we

Amino acids and Proteins in Fossil Biominerals: An Introduction for Archaeologists and Palaeontologists, First Edition. Beatrice Demarchi.
© 2020 John Wiley & Sons Ltd. Published 2020 by John Wiley & Sons Ltd.

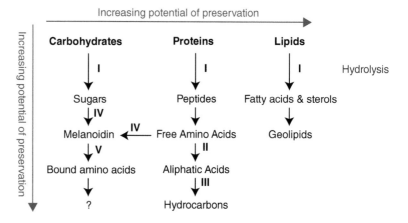

Scheme 2.1 Hoering's summary of diagenesis pathways affecting organic matter in fossil mollusc shells (Hoering, 1980).

attempt to describe diagenesis in mathematical terms? We usually do so when we wish to extrapolate the extent of diagenesis as a function of time – generally for dating purposes, though also as a means of predicting the presence of intact biomolecules in old fossils. In the case of protein diagenesis dating, the observed variable will be one of the diagenetic indicators (typically, the extent of racemization of a certain amino acid), which is determined for a range of samples of a known age. Using a 'black box' approach (Kriausakul and Mitterer, 1980; Wehmiller, 1980; Collins and Riley, 2000), we can describe the relationship between extent of racemization and time using a mathematical function; this has the obvious advantage that the same function can then be used to extrapolate the age of a sample on the basis of the measured extent of racemization. However, one should always remember that the observed variable is the combination of:

- all the processes (e.g. racemization, hydrolysis, decomposition, condensation) of
- all the (thousands of) amino acids in
- all of the (hundreds of) proteins
- that are degrading at different rates
- while trapped together with other organics (with which they interact)
- within a nanometric 'void' of random shapes and positions
- within a mineral crystal (with one of many chemical formulae and crystallographic parameters, as in Table 1.1, Chapter 1)
- next to other mineral crystals
- that form a specific tissue
- that is buried in a specific environment.

These multiple organic–organic and organic–inorganic interfaces and interactions characterize every aspect of the processes of diagenesis, and make the understanding of the main mechanisms of degradation (and

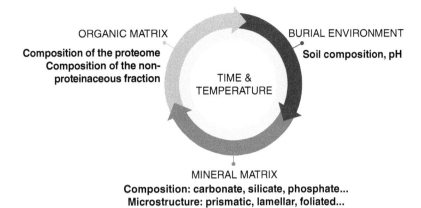

ORGANIC MATRIX
Composition of the proteome
Composition of the non-
proteinaceous fraction

BURIAL ENVIRONMENT
Soil composition, pH

TIME &
TEMPERATURE

MINERAL MATRIX
Composition: carbonate, silicate, phosphate...
Microstructure: prismatic, lamellar, foliated...

Scheme 2.2 A summary of the main factors affecting diagenesis.

survival) very challenging. Therefore, it becomes necessary to artificially 'remove' some of this complexity in order to pinpoint some of the pathways of diagenesis. The main factors influencing diagenesis, and strategies for circumventing their effect, are listed here. These factors are also summarized in Scheme 2.2.

- The burial environment, including soil composition and pH: its effect on protein decay can be circumvented by isolating the intracrystalline fraction only (by bleaching or as a by-product of diagenesis). If this behaves as a 'closed system' the only environmental factor influencing diagenesis (other than time) is the temperature experienced by the sample throughout its life and post mortem (see Towe and Thompson, 1972; Sykes et al., 1995; Penkman et al., 2008, for example).
- The chemical composition of the mineral (e.g. carbonate versus silica), crystal morphology (e.g. aragonite versus calcite), microstructural arrangement (e.g. platelets versus prisms): these factors can only be normalized by comparing the same type of biominerals, down to the level of genus/species. Furthermore, some biominerals, such as mollusc shell, have both calcitic and aragonitic layers, with different microstructures, therefore each layer should be considered separately (Hearty et al., 1986; Demarchi et al., 2013a; Torres et al., 2013).
- The chemical composition of the organic matrix; very little is known about the variability of the nonproteinaceous component of the organic matrix in biominerals, therefore the assumption must be made that comparing similar with similar (e.g. microstructural layers, as above) will be sufficient to minimize this source of variability.
- The composition of the biomineral proteome, which includes hundreds of proteins (e.g. 273 proteins confidently identified in modern bleached ostrich eggshell: Demarchi et al., 2016): this factor is crucial, because the degradation of each amino acid is influenced by the

chemical and steric characteristics of its neighbouring residues, as well as by the protein conformation, which will be progressively lost as diagenesis proceeds. Furthermore, the occlusion of proteins in the crystals may occur randomly as the crystals grow around them. In the case of rapidly mineralizing systems, such as avian eggshell, this may result in high variability in the type and proportion of proteins trapped. Finally, direct interaction with the mineral surface may result in unexpected patterns of survival (Demarchi et al., 2016). The compositional variability of the proteome can be partially controlled by analysing several biological replicates.

- The combined effect of time and temperature in the burial environment is also difficult to predict, because it depends on altitude, burial depth, type of soil, presence or absence of vegetation, moisture, exposure to direct sunlight, erosion and reburial (Wehmiller, 1977; Miller et al., 1992; Collins and Demarchi, 2015): the effect of this can be elucidated by performing artificial diagenesis experiments, in which a fragment of biomineral (often powdered to a certain grain size, bleached or unbleached) is immersed in water (or moist sand) in sealed glass ampoules and heated at high temperatures for known times (Hare and Mitterer, 1969; Brooks et al., 1991; Goodfriend and Meyer, 1991; Penkman et al., 2008). This is very useful for determining some of the degradation patterns at high temperature, but the resulting data must be used with caution when attempting to explain patterns at the normal (low) burial temperature, both because the temperature sensitivities of the vast range of reactions involved are such that the order of processes will be affected, and also because the mineral phase will be affected differently by high and low temperatures (Demarchi et al., 2013b; Tomiak et al., 2013).

Overall, it is evident that considering protein diagenesis 'in isolation' is problematic. However, understanding the main chemical mechanisms that drive breakdown is important for many applications of ancient protein studies, from dating to reconstructing and authenticating ancient sequences. Therefore, here we will briefly describe three main mechanisms of decay:

1 hydrolysis of the peptide bonds
2 racemization of the individual amino acids (either peptide bound or free)
3 decomposition of amino acids (and of their degradation products).

However, we must bear in mind that there are many other diagenesis-induced modifications, and these are frequently encountered in ancient protein sequences, for example, the deamidation of Asn and Gln, the oxidation of Met and Trp, and dehydration processes (sometimes followed by formation of compounds such as pyroglutamic acid). These diagenesis-induced modifications, deamidation in particular, can offer precious insights into protein sequence authenticity and preservation (Demarchi et al., 2016; Welker et al., 2016; Mackie et al., 2018).

2.2 Hydrolysis

Proteins are synthesized in the ribosomes, where individual amino acids are put together via peptide bonds between neighbouring residues. This condensation reaction between the $-NH_2$ and the $-COOH$ groups leads to the loss of a water molecule. The peptide bond is the weakest bond in the primary structure of a protein and it can be lysed by the addition of a molecule of water, that is, via hydrolysis (Figure 2.1).

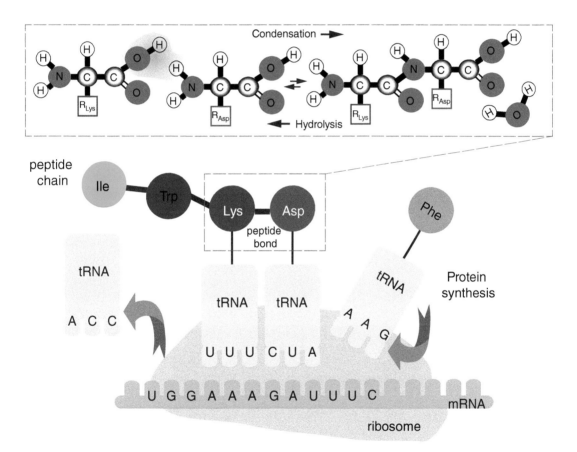

Figure 2.1 Protein synthesis in the ribosomes of living organisms, involving tRNAs and mRNA molecules. Peptide bonds are formed by condensation of the carboxy and amino groups of two neighbouring residues (Asp and Lys in this example) and a molecule of water is released every time an amino acid is added to the growing peptide chain. The reverse reaction, hydrolysis, requires one molecule of water per peptide bond and occurs during protein breakdown. Image sources: https://commons.wikimedia.org/wiki/File:Peptidformationball_uk.svg; https://commons.wikimedia.org/wiki/File:Protein_Synthesis-Translation.png.

Hydrolysis is thus the most likely mechanism to drive early protein diagenesis. In most cases, it appears that hydrolysis precedes racemization and decomposition: artificial degradation experiments have shown that the activation energy for hydrolysis is lower than for other breakdown reactions (Collins and Riley, 2000). Obviously, water is required: this is present within the intracrystalline fraction of biominerals as fluid inclusions (Hudson, 1967; Gaffey, 1988) but also as protein structural water (Bellissent-Funel et al., 2016) and as a chemical by-product of decomposition and condensation reactions (Bada et al., 1978). However, this pool of water may eventually run out, preventing further hydrolysis of the peptide bonds.

The rates of hydrolysis by random peptide bond scission events in a closed system should theoretically conform to first-order irreversible kinetics (Collins and Riley, 2000):

$$-\ln[1 - \sqrt{\gamma_1}] = kt$$

where γ is the weight fraction of free amino acids (FAA) at a certain time (t) and k is the rate constant for hydrolysis of a polypeptide with a number of residues greater than 16. The error generated by the use of this equation decreases proportionally to the number of initial residues, and it will be 1% when the chain includes more than 50 residues (Collins and Riley, 2000). γ can be quantified by measuring the concentrations of both the free amino acids (FAA) and the total hydrolysable amino acids (THAA: free + peptide bound) (see Chapter 4), as follows:

$$\%FAA = \frac{[FAA]}{[THAA]} \times 100$$

If hydrolysis occurs by random scission, then the plot of $-\ln[1 - \sqrt{\gamma}]$ versus time should yield a straight line with slope k (the dashed red line in Figure 2.2). However, data on closed-system hydrolysis of proteins from ostrich eggshell (Miller et al., 1992) and mollusc shells (*Patella vulgata* among others, Figure 2.2) do not conform to this model: at least two breaks of slope can be observed. Therefore there are different rates of reaction, with very slow hydrolysis dominating the last stages of diagenesis. There are three possible causes for this:

1 the accumulation of a (hydrophobic) hydrolysis-resistant fraction
2 the effect of surface stabilization, which results in a much higher energetic barrier for hydrolytic scission
3 the system simply runs out of water.

These three possible causes are now discussed.

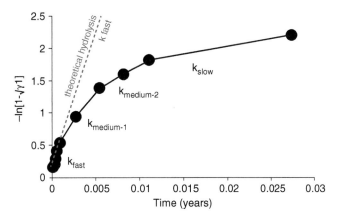

Figure 2.2 Peptide bond hydrolysis should conform to first-order irreversible kinetics if it was a series of random chain-scission events (e.g. red dashed line). However, real data obtained by measuring the extent of hydrolysis in closed-system biominerals (bleached heated *Patella* shell powders in this example) show a progressive slowing down of the reaction rate. The presence of a hydrolysis-resistant fraction is one of the possible causes for this effect. This fraction may be highly hydrophobic, or it may be hydrophilic but stabilized by the mineral. Furthermore, the water in the system (the nanometric voids) may eventually run out.

2.2.1 The Accumulation of a Hydrolysis-resistant Fraction

The probability of peptide bond lysis is not equal for all bonds in the macromolecule: hotspots for hydrolysis within the protein chain (internal hydrolysis) under biological conditions (mild pH and temperature) will be those peptide bonds involving hydrophilic amino acids, i.e. those with a polar or electrically charged side-chain (see Box 1.1 in Chapter 1). On the contrary, bonds between aliphatic and bulky amino acids should be harder to break. Over time we should observe the accumulation of a residual hydrolysis-resistant peptide fraction, which is aliphatic in nature. Indeed, a fraction of amino acids bound in compounds that are hydrolysis resistant (i.e. cannot be broken down even with harsh acid hydrolysis at high temperatures) have been detected experimentally in ostracods, mollusc shells and ostrich eggshell (Bright and Kaufman, 2011; Crisp et al., 2013; Demarchi et al., 2013a). Hoering (1980) suggested that interactions between the reducing groups of sugars and amino groups of other compounds form glycosylamines, with the end product of the reaction being a dark heteropolymer referred to as a melanoidin: a compound which shares many properties of natural humic acids found in soils and sediments. Interestingly, these compounds can liberate free amino acids upon hydrolysis, but at a different rate to that of peptide hydrolysis (Hoering, 1980;

Collins et al., 1992; Walton, 1998). The role of condensation between proteins and nonproteinaceous components of the intracrystalline system is one that ought to be investigated in greater detail, but has been skirting at the edge of consciousness of palaeobiogeochemists for a long time without ever being systematically addressed. This should be rectified soon, because any attempts at modelling hydrolysis data obtained experimentally clash against the rock of the hydrolysis-resistant fraction by removing available amino acid residues from the equation. One possibility is to use a long time series of monospecific samples with simple microstructures and assess the relative composition of the system, with regard to the 'hydrolysable' and 'nonhydrolysable' fractions, using a combination of analytical techniques. This has been partially attempted for ostrich eggshell (see Crisp et al., 2013; Demarchi et al., 2016), but the characterization and quantification of compounds of unknown chemical composition is extremely complicated.

2.2.2 *The Effect of Surface Stabilization*

The assumption that bonds between hydrophilic amino acids are the weakest (Hill, 1965) is reasonable, and as such we do not expect to find many peptide-bound hydrophilic amino acids in old fossil biominerals. However, the opposite appears to be true, at least in some biomineralized systems: the oldest authenticated peptide sequence reported to date, extracted from ca. 3.8 Ma old ostrich eggshell (Laetoli, Tanzania), contains four Asp residues (Demarchi et al., 2016): in this case, the hypothesis (which was supported by computational modelling) is that the negatively charged Asp binds strongly to the positive calcium ions of the calcite surface. However, water is present nearby (the 'layered' or 'structured' water, Figure 2.3A), which should easily break the Asp–Asp bonds. These are highly ordered water molecules, stabilized by the mineral surface. This water is somewhat 'frozen' because of its highly ordered nature, and cannot participate in peptide bond degradation. In addition, the Asp–Ca binding causes an increase in entropy (disrupting the layered water) and therefore results in an even stronger binding. Overall, the effect of surface stabilization is such that the energy barrier required for hydrolysis in solution (i.e. in absence of mineral surfaces) is lower than that obtained when the protein is attached to the mineral surface (Figure 2.3B). The mechanistic study on ostrich eggshell has therefore elucidated a pathway for preservation which has long been suggested for archaeological and palaeontological bone (Collins et al., 2000; San Antonio et al., 2011), mollusc shell (Weiner, 1983), and for sediments in shallow marine carbonate environments (Cunningham and Mitterer, 1980): a combination of the protein's structural properties and its 'intimate association' with the mineral (which is often invoked but seldom clarified at the molecular level).

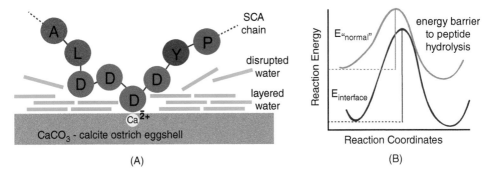

(A) (B)

Figure 2.3 (A) Schematic representation of the Asp-rich domain of struthiocalcin-1 (the dominant protein in ostrich eggshell), able to bind to the Ca^{2+} of the calcite surface by disrupting the layers of ordered water at the interface. The remaining layered water is 'frozen' in position and unavailable to break the peptide bonds. (B) The net effect of the increase in entropy due to the disruption and the unavailability of the layered water is strong binding between Asp and Ca^{2+}, and a higher than normal energy barrier for hydrolysis of the Asp–Asp bonds, which are therefore preserved over geological timescales. Figures derived from Demarchi et al. (2016) and Wallace and Schiffbauer (2016).

2.2.3 Lack of Water

The intracrystalline proteins are occluded within nanometric voids: the amount of water present in these voids is extremely difficult to estimate. Limitation of water available for peptide bond hydrolysis may be a factor: water present as fluid inclusions may be consumed during early diagenesis, while water layered next to the mineral surface is unavailable (see Section 2.2.2). Water is also produced by decomposition reactions of some amino acids (Ser, Thr and Glu), but the importance of this 'diagenetic water' in driving hydrolysis forward is still unclear.

In general, it is clear that absence of water for peptide hydrolysis is a key factor for protein preservation in closed systems occluded in biominerals, but whether water is not physically present or is simply unavailable for the reaction needs to be investigated on a case-by-case basis.

2.3 Racemization

Racemization, the interconversion between two chiral forms (stereoisomers or enantiomers) of the same amino acid, is the key diagenetic reaction for geochronological applications. Chirality (from the Greek 'χείρ', hand), is a property of any amino acid that has four different groups bound to the central carbon atom and is therefore asymmetric (Figure 2.4); glycine is an exception because its –R group is an –H (and is therefore symmetric; see Box 1.1 in Chapter 1), while isoleucine, hydroxyproline, hydroxylysine and

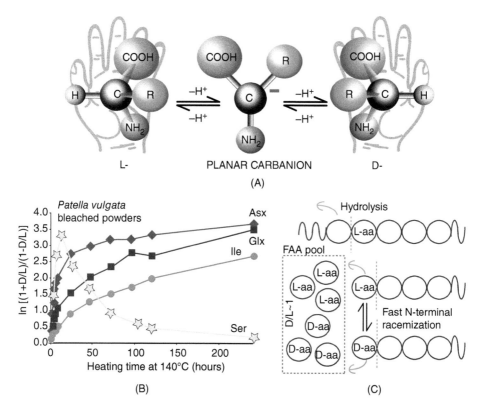

Figure 2.4 (A) Chirality and mechanism of racemization via a planar carbanion (Neuberger, 1948); drawing adapted from https://commons.wikimedia.org/wiki/File:Chirality_with_hands.svg. (B) High-temperature data (artificial diagenesis experiments) obtained on the gastropod *Patella vulgata* (bleached powder) showing nonlinear racemization for Asx (Asp + Asn), Glx (Glu + Gln) and Ile; note the different racemization rates of the three amino acids (data from Demarchi et al., 2013b). (C) Schematic of N-terminal peptide bond hydrolysis, releasing highly racemized AAs in the free pool (modified after Demarchi and Collins, 2015).

threonine have two chiral centres (and therefore four stereoisomers, called diastereomers). The interconversion between diastereoisomers is more properly called epimerization, but the term racemization is used routinely for both reactions and hence that terminology is also used here. The enantiomers are mirror images, but cannot be superimposed, just like human hands – and the simplest nomenclature of the enantiomers (the one used in palaeobiogeochemistry) derives from this property: the L and D forms are named *laevo* and *dextro* ('left' and 'right' in Latin).

The amino acids used for protein synthesis (see Figure 2.1) are L enantiomers (exceptions are D-amino acids in bacterial peptidoglycans: Schleifer and Kandler, 1972); the interconversion from L to D occurs post mortem only (or in the case of lack of tissue turnover in vivo, for example in the eye lens) and it is thermodynamically favoured as it brings a system from a state of disequilibrium towards equilibrium. The mechanism for racemization of free amino acids in aqueous solution was put forward in the mid-20th century by Neuberger (1948) and has not been updated since. Neuberger proposed that the key step was the abstraction of the hydrogen atom linked to the central carbon atom, which causes the collapse of the tetrahedral geometry of the amino acid and the formation of a planar carbanion. The H^+ can then add back to either side of the planar carbanion, generating either an L or a D form (Figure 2.4A). The probability of the abstraction will depend on a variety of factors, including the chemical and steric characteristics of the side-chain of the amino acid itself, which can either stabilize or destabilize the planar carbanion, and therefore some amino acids will racemize faster than others.

The reaction should theoretically conform to first-order reversible kinetics for free amino acids in aqueous solution (Bada and Schroeder, 1972):

$$\text{L} \underset{k_D}{\overset{k_L}{\rightleftharpoons}} \text{D}$$

where L and D represent the concentration of the two enantiomers and k_L and k_D are the forward and backward rate constants. As the reaction proceeds, the D forms increase in concentration until an equilibrium mixture of D and L enantiomers is achieved (racemic mixture). The ratio between the D and L forms is called the DL ratio or D/L value, and it depends upon the ratio between the forward and backward rate constants. For amino acids with a single chiral centre the DL ratio at equilibrium will be 1:1 ($k_L = k_D$), whilst for multi-asymmetric amino acids the equilibrium ratio may differ significantly from 1 (the k_L/k_D ratio for L-isoleucine to D-alloisoleucine has been found to be about 1.3:1) (Bada, 1985).

The derived first-order equation is:

$$\ln \left(\frac{1 + \dfrac{D}{L}}{1 - \dfrac{D}{L}} \right) - \text{constant} = 2kt$$

where t is time and the constant term represents the small amount of D enantiomer already naturally present in modern samples and the small amount of racemization induced by the hydrolysis step during sample preparation.

For amino acids with more than one chiral centre a modified kinetic equation is required. For isoleucine, for example:

$$\ln \left(\frac{1 + \dfrac{A}{I}}{1 - \dfrac{K'A}{I}} \right) - \text{constant} = \left(1 + K'\right)k_{\text{Ile}}t$$

where $K' = 1/K_{eq}$ and K_{eq} is the alloisoleucine/isoleucine (A_{eq}/I_{eq}) ratio at equilibrium, which is ~1.3 to 1.4 (Wehmiller and Hare, 1971; Bada and Schroeder, 1972, 1975).

While this may be truly representative of the racemization of free amino acids in solution, it has very little bearing on racemization in biominerals. Experimental data on amino acid racemization in biominerals (both from high-temperature artificial diagenesis experiments and from real fossil biominerals of known age) clearly deviate from the first-order pattern: a plot of $\ln[(1 + D/L)/(1 - D/L)]$ versus time does not yield a simple straight line (Figure 2.4B). There are many reasons for this, but perhaps the most important is that, in most cases, hydrolysis drives racemization (as we will discuss below) and therefore all the complicating factors which affect hydrolysis rates also impact directly on racemization.

Hydrolysis influences the state of each amino acid (Figure 2.4C): free or peptide-bound, internal or terminal (and, in the latter case, N- or C-terminal). As a general rule, loss of the hydrogen is highly unlikely for amino acids bound in the peptide chain, therefore in-chain racemization does not usually occur. On the contrary, an N-terminal amino acid will racemize easily, because of the electron-withdrawing and resonance-stabilizing characteristics of the N terminus, which facilitates –H loss; this results in a high probability of a D enantiomer being released in the free pool (Figure 2.4C). C-terminal racemization is only enhanced in the case of diketopiperazine (DKP) formation (Steinberg and Bada, 1981, 1983; Fischer, 2003). These are the smallest cyclic peptides, which are preferentially formed when Gly occupies the third position from the amino terminus in a protein sequence (Sepetov et al., 1991). Hydrolysis of DKPs may yield the original dipeptide or an inverted dipeptide. Moreover, racemization within the cyclic structure is extremely rapid at neutral pH (Steinberg and Bada, 1981), therefore causing anomalously high DL ratios, even greater than 1.5, as observed for model tetrapeptides by Moir and Crawford (1988).

There are, of course, exceptions to this rule: Asp, Asn (collectively called 'Asx' to indicate the combined analytical signal of Asn and Asp: during sample preparation the acid hydrolysis step causes the decomposition of asparagine, Asn, into Asp: Hill, 1965), as well as Ser, can undergo in-chain racemization. Asx degradation is thought to occur via the formation of a five-membered succinimide ring (Asu) via intramolecular cyclization in which the α-amino group of the carboxyl amino acid residue attacks the side-chain carbonyl carbon of an aspartyl/asparaginyl residue (Collins et al., 1999). This mechanism contributes to deamidation, isomerization and racemization of the residue. In particular, racemization within the succinimide ring is enhanced by resonance stabilization of the carbanion by both the α-carbonyl and the β- carbonyl (Geiger and Clarke, 1987): L-Asp is rapidly converted to L-Asu, which racemizes to D-Asu and is then transformed to D-Asp. Asu racemization can be five orders of magnitude faster than free Asp (Radkiewicz et al., 1996). In-chain Ser racemization has been observed experimentally in a kinetic study of a model peptide (Demarchi et al.,

2013c), but the mechanisms have yet to be fully clarified (Takahashi et al., 2010, 2017).

The patterns of racemization are highly complex, even in simple model systems (e.g. synthetic peptide in solution, degraded at known temperatures for known times), therefore it is somewhat surprising that it is possible to use this reaction effectively for geochronological purposes, as this requires a definite relationship between the extent of racemization (D/L) and time. AAR dating requires that racemization follows *predictable* patterns (which may be described mathematically), even if we may not understand the mechanisms underlying this.

2.4 Decomposition and Other Diagenesis-induced Modifications

It is well known that amino acids (free or bound) can decompose into simpler constituents, following a variety of reaction pathways (Table 2.1). The experimental proof of this has been obtained using artificial diagenesis experiments: there is a quantifiable loss of amino acids from the intracrystalline fraction, which cannot be ascribed to leaching from the system (when this fraction approximates a closed system) (Brooks et al., 1990; Penkman et al., 2008; Bright and Kaufman, 2011; Orem and Kaufman, 2011; Hendy et al., 2012; Crisp et al., 2013; Demarchi et al., 2013a, 2015; Ortiz et al., 2015; Pierini et al., 2016).

Table 2.1 The most common reaction products of individual amino acids at high temperature and in fossil biominerals. Data summarized from Vallentyne (1964), Bada (1991) and Walton (1998).

Amino acid	Process	Product
Leu, Val, Phe, Gly, Ser, Thr, Ala	Decarboxylation	Amine + CO_2 e.g. Ala → CO_2 + ethylamine
Asp	α,β-decarboxylation	β-Ala[1] or Ala
Glu	γ-decarboxylation	γ-aminobutyric acid[1]
Free Glu	Reversible lactamization	Pyroglutamic acid
Asp	Reversible deamination	Fumaric acid (+ NH_3)
Asn, Gln	Irreversible deamidation	Asp, Glu + NH_3
Ser	Dehydration	Racemic Ala
Thr	Dehydration	Racemic α-aminobutyric acid
Ser, Thr	Aldol cleavage	Gly + aldehydes
Arg		Urea+ ornithine

[1] Neither β-Ala nor γ-aminobutyric acid have been observed in foraminifera (Schroeder, 1975) or in heated solutions of Asp (Bada and Miller, 1970), although they are extremely abundant in sediments (Schroeder and Bada, 1976) and are therefore thought to derive from the biological degradation of Asp and Glu (Vallentyne, 1964; Bada, 1991; Walton, 1998).

Dehydration of Ser to Ala is the reaction that has traditionally been studied the most, due to its geochronological applications. The work of Bada et al. (1978) and Bada and Man (1980) on foraminifera, and further studies carried out on the intracrystalline proteins in shells, corals, eggshells and opercula, as well as on model peptides, have shown that the ratio of the concentration of Ser to Ala (the [Ser]/[Ala] value) decreases with time following predictable patterns (Penkman et al., 2008, 2011; Demarchi et al., 2011; Davies et al., 2012). These can be described by an exponential relationship for some biominerals (e.g. *Bithynia* opercula data used by Westaway, 2009) but in most instances the reaction cannot readily be modelled using mathematical functions. The interplay of Ser decomposition and in-chain racemization is complex and results in (a) Ser D/L values higher than 1, reached extremely rapidly and (b) reversal of D/L values with time, rather than the plateau attained by all other amino acids at equilibrium (see Figure 2.4B).

Deamidation of Asn and Gln to Asp and Glu respectively is one of the best-known diagenetic processes. Deamidation occurs via a cyclic intermediate (glutarimide/succinimide) at neutral or alkaline pH or via direct hydrolysis of the side-chain at low pH or in presence of metal ions. The mechanisms have been elucidated for Asn (Bada and Miller, 1970; Capasso et al., 1993) and more recently for Gln (Li et al., 2010) but, whatever the chemical pathways followed, the end result is a mass 'gain' of +0.98402 Da, because an amide side-chain group is converted to a carboxylate. Since the rates of Asn deamidation are fast, this reaction is a useful marker for mild degradation of proteins, while Gln deamidation is slower and its potential for palaeontological and archaeological applications higher. In the last few years, Gln deamidation has been used as a semi-quantitative marker of peptide degradation (Leo et al., 2011) and therefore of the antiquity and authenticity of ancient protein sequences (van Doorn et al., 2012; Wilson et al., 2012; Welker et al., 2015). However, sample preparation, as well as a plethora of environmental and experimental factors, influences Gln deamidation rates (Simpson et al., 2016, 2019). As such, the extent of Gln deamidation is better considered as an indicator of preservational quality on a sample-specific basis (Schroeter and Cleland, 2016).

In general, racemization rates in nonclosed systems are susceptible to large degrees of variation, due to changes within the burial environment over time, and the same applies to Gln deamidation and other diagenesis-induced modifications. Palaeobiogeochemical studies have repeatedly shown that any attempt at comparing different fossils, protein fractions and burial environments is unlikely to succeed. On the contrary, if we consider a series of biominerals of increasing age, which contain a closed-system fraction of proteins, then the patterns observed are meaningful. Figure 2.5 shows the degradation of the intracrystalline protein struthiocalcin-1 isolated from ostrich eggshell samples from a range of sites in South Africa

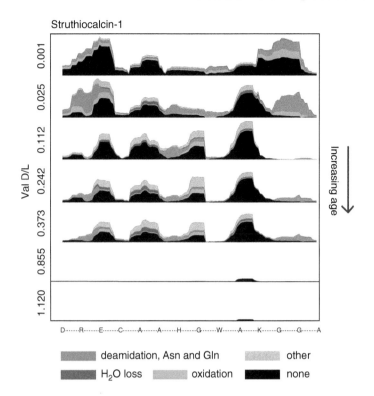

Figure 2.5 Degradation of struthiocalcin-1, the dominant protein of the intracrystalline fraction within ostrich eggshell, measured in a series of fossil samples from Africa (Elands Bay Cave and Pinnacle Point, South Africa; Laetoli, Tanzania). Over time (and increasing THAA Val D/L values) degradation progresses and the count of identified peptides diminishes. Facile diagenesis reactions such as Asn and Gln deamidation proceed rapidly during early diagenesis (e.g. D/L values of 0.001–0.025) and their extent increases; however the apparent extent of deamidation decreases as the peptide coverage is progressively lost. Adapted from Demarchi et al. (2016).

(with the exception of the oldest, which comes from Tanzania: Demarchi et al., 2016). It is clear that:

a the extent of diagenesis-induced modifications (among which we highlighted deamidation, water loss, and oxidation) increases steadily with increasing age

b the overall sequence preservation for each region of the protein decreases with increasing age, as shown by the number of peptides identified by mass spectrometry.

As a consequence, the extent of deamidation, for example, will also decrease as the direct consequence of overall degradation – this will result in variable extents of deamidation, which leads to complications when trying to correlate the extent of Gln deamidation to the chronometric age of the protein.

2.4.1 A Note on Authentication

This brief review has shown that there is a great variety of chemical reactions which simultaneously transform and degrade the organic matter within biominerals, even when this occurs within a *closed system*, in which, apart from temperature, no environmental factors related to the time of burial are considered. Microbial and fungal degradation, both of which are extremely important players in the decomposition of all organic matter, are thus ignored.

It follows that a protein that is genuinely ancient must be degraded in some way – there is no known process capable of halting degradation. This means that, for any proteins which have been recovered from subfossil biominerals (or any other system) and which are intact and pristine, there is an extremely high chance that these proteins have been subjected to modern contamination. It is crucial that data on the degradation of proteins be included in any and all studies dealing with ancient molecules (Hendy et al., 2018).

References

Bada, J.L. (1985). Racemization of amino acids. In: *Chemistry and Biochemistry of the Amino Acids* (ed. G.C. Barrett), 399–414. Springer.

Bada, J.L. (1991). Amino acid cosmogeochemistry. *Philos. Trans. R. Soc. Lond. B Biol. Sci.*, 333: 349–358.

Bada, J.L. and Man, E.H. (1980). Amino acid diagenesis in Deep Sea Drilling Project cores: Kinetics and mechanisms of some reactions and their applications in geochronology and in paleotemperature and heat flow determinations. *Earth-Sci. Rev.*, 16: 21–55.

Bada, J.L. and Miller, S.L. (1970). Kinetics and mechanism of the reversible nonenzymic deamination of aspartic acid. *J. Am. Chem. Soc.*, 92: 2774–2782.

Bada, J.L. and Schroeder, R.A. (1972). Racemization of isoleucine in calcareous marine sediments: Kinetics and mechanism. *Earth Planet. Sci. Lett.*, 15: 1–11.

Bada, J.L. and Schroeder, R.A. (1975). Amino acid racemization reactions and their geochemical implications. *Naturwissenschaften*, 62: 71–79.

Bada, J.L., Shou, M.-Y., Man, E.H., and Schroeder, R.A. (1978). Decomposition of hydroxy amino acids in foraminiferal tests; kinetics, mechanism and geochronological implications. *Earth Planet. Sci. Lett.*, 41: 67–76.

Bellissent-Funel, M.-C., Hassanali, A., Havenith, M. et al. (2016). Water determines the structure and dynamics of proteins. *Chem. Rev.*, 116: 7673–7697.

Bright, J. and Kaufman, D.S. (2011). Amino acid racemization in lacustrine ostracodes, part I: Effect of oxidizing pre-treatments on amino acid composition. *Quat. Geochronol.*, 6: 154–173.

Brooks, A.S., Hare, P.E., Kokis, J.E. et al. (1990). Dating Pleistocene archaeological sites by protein diagenesis in ostrich eggshell. *Science*, 248: 60–64.

Brooks, A.S., Hare, P.E., Kokis, J.E., and Durana, K. (1991). A burning question: Differences in laboratory induced and natural diagenesis in ostrich eggshell proteins. *Annu. Rep. Geophys. Lab., Carnegie Institution of Washington*, 176–179.

Capasso, S., Mazzarella, L., Sica, F. et al. (1993). Kinetics and mechanism of succinimide ring formation in the deamidation process of asparagine residues. *J. Chem. Soc., Perkin Trans.*, 2: 679–682.

Collins, M. and Demarchi, B. (2015). Amino acid racemization, paleoclimate. In: *Encyclopedia of Scientific Dating Methods*, 47-48. Springer.

Collins, M.J. and Riley, M.S. (2000). Amino acid racemization in biominerals: The impact of protein degradation and loss. In: *Perspectives in Amino Acid and Protein Geochemistry* (ed. G. Goodfriend, M.J. Collins., M. Fogel et al.), 120–141. Oxford University Press.

Collins, M.J., Westbroek, P., Muyzer, G., and de Leeuw, J.W. (1992). Experimental evidence for condensation reactions between sugars and proteins in carbonate skeletons. *Geochim. Cosmochim. Acta*, 56: 1539–1544.

Collins, M.J., Waite, E.R., and van Duin, A.C. (1999). Predicting protein decomposition: The case of aspartic-acid racemization kinetics. *Philos. Trans. R. Soc. Lond. B Biol. Sci.*, 354: 51–64.

Collins, M.J., Gernaey, A.M., Nielsen-Marsh, C.M. et al. (2000). Slow rates of degradation of osteocalcin: Green light for fossil bone protein? *Geology*, 28: 1139–1142.

Crisp, M., Demarchi, B., Collins, M. et al. (2013). Isolation of the intra-crystalline proteins and kinetic studies in *Struthio camelus* (ostrich) eggshell for amino acid geochronology. *Quat. Geochronol.*, 16: 110–128.

Cunningham, R. and Mitterer, R.M. (1980). Metal binding study of fulvic acids from carbonate sediments using manganese as a magnetic resonance probe. In: *Biogeochemistry of Amino Acids* (eds. P.E. Hare, T.C. Hoering and K. King Jr.), 129–143. New York: Wiley.

Davies, B.J., Roberts, D.H., Bridgland, D.R. et al. (2012). Timing and depositional environments of a Middle Pleistocene glaciation of northeast England: New evidence from Warren House Gill, County Durham. *Quat. Sci. Rev.*, 44: 180–212.

Demarchi, B. and Collins, M. (2015). Amino acid racemization dating. In: *Encyclopedia of Scientific Dating Methods*, 13-26. Springer.

Demarchi, B., Williams, M.G., Milner, N. et al. (2011). Amino acid racemization dating of marine shells: A mound of possibilities. *Quat. Int.*, 239: 114–124.

Demarchi, B., Rogers, K., Fa, D.A. et al. (2013a). Intra-crystalline protein diagenesis (IcPD) in *Patella vulgata*. Part I: Isolation and testing of the closed system. *Quat. Geochronol.*, 16: 144–157.

Demarchi, B., Collins, M.J., Tomiak, P.J. et al. (2013b). Intra-crystalline protein diagenesis (IcPD) in *Patella vulgata*. Part II: Breakdown and temperature sensitivity. *Quat. Geochronol.*, 16: 158–172.

Demarchi, B., Collins, M., Bergström, E. et al. (2013c). New experimental evidence for in-chain amino acid racemization of serine in a model peptide. *Anal. Chem.*, 85: 5835–5842.

Demarchi, B., Clements, E., Coltorti, M. et al. (2015). Testing the effect of bleaching on the bivalve *Glycymeris*: A case study of amino acid geochronology on key Mediterranean raised beach deposits. *Quat. Geochronol.*, 25: 49–65.

Demarchi, B., Hall, S., Roncal-Herrero, T. et al. (2016). Protein sequences bound to mineral surfaces persist into deep time. *eLife*, **5**: e17092.

Fischer, P.M. (2003). Diketopiperazines in peptide and combinatorial chemistry. *J. Pept. Sci.*, 9: 9–35.

Gaffey, S.J. (1988). Water in skeletal carbonates. *J. Sediment. Res.*, 58: 397–414.

Geiger, T. and Clarke, S. (1987). Deamidation, isomerization, and racemization at asparaginyl and aspartyl residues in peptides. Succinimide-linked reactions that contribute to protein degradation. *J. Biol. Chem.*, 262: 785–794.

Goodfriend, G.A. and Meyer, V.R. (1991). A comparative study of the kinetics of amino acid racemization/epimerization in fossil and modern mollusk shells. *Geochim. Cosmochim. Acta*, 55: 3355–3367.

Hare, P.E. and Mitterer, R.M. (1969). Laboratory simulation of amino acid diagenesis in fossils. *Carnegie Institution of Washington Yearbook*, 67: 205–208.

Hearty, P.J., Miller, G.H., Stearns, C.E., and Szabo, B.J. (1986). Aminostratigraphy of Quaternary shorelines in the Mediterranean basin. *Geol. Soc. Am. Bull.*, 97: 850–858.

Hendy, E.J., Tomiak, P.J., Collins, M.J. et al. (2012). Assessing amino acid racemization variability in coral intra-crystalline protein for geochronological applications. *Geochim. Cosmochim. Acta*, 86: 338–353.

Hendy, J., Welker, F., Demarchi, B. et al. (2018). A guide to ancient protein studies. *Nat. Ecol. Evol.*, 2: 791–799.

Hill, R.L. (1965). Hydrolysis of proteins. *Adv. Protein Chem.*, 20: 37–107.

Hoering, T.C. (1980). The organic constituents of fossil mollusk shells. In: *Biogeochemistry of Amino Acids* (eds. P.E. Hare, T.C. Hoering and K. King Jr.), 193–201. New York: Wiley.

Hudson, J.D. (1967). The elemental composition of the organic fraction, and the water content, of some recent and fossil mollusc shells. *Geochim. Cosmochim. Acta*, 31: 2361–2378.

Kriausakul, N. and Mitterer, R.M. (1980). Some factors affecting the epimerization of isoleucine in peptides and proteins. In: *Biogeochemistry of Amino Acids* (eds. P.E. Hare, T.C. Hoering and K. King Jr.), 283–296. New York: Wiley.

Leo, G., Bonaduce, I., Andreotti, A. et al. (2011). Deamidation at asparagine and glutamine as a major modification upon deterioration/aging of proteinaceous binders in mural paintings. *Anal. Chem.*, 83: 2056–2064.

Li, X., Lin, C., and O'Connor, P.B. (2010). Glutamine deamidation: Differentiation of glutamic acid and γ-glutamic acid in peptides by electron capture dissociation. *Anal. Chem.*, 82: 3606–3615.

Mackie, M., Rüther, P., Samodova, D. et al. (2018). Palaeoproteomic profiling of conservation layers on a 14th century Italian wall painting. *Angew. Chem. Int. Ed. Engl.*, 57: 7369–7374.

Miller, G.H., Beaumont, P.B., Jull, A.J.T. et al. (1992). Pleistocene geochronology and palaeothermometry from protein diagenesis in ostrich eggshells: Implications for the evolution of modern humans. *Phil. Trans. R. Soc. Lond. B*, **337**: 149–157.

Moir, M.E. and Crawford, R.J. (1988). Model studies of competing hydrolysis and epimerization of some tetrapeptides of interest in amino acid racemization studies in geochronology. *Can. J. Chem.*, 66: 2903–2913.

Neuberger, A. (1948). Stereochemistry of amino acids. In *Advances in Protein Chemistry*, vol. 4, 297–383. Elsevier.

Orem, C.A. and Kaufman, D.S. (2011). Effects of basic pH on amino acid racemization and leaching in freshwater mollusk shell. *Quat. Geochronol.*, 6: 233–245.

Ortiz, J.E., Gutiérrez-Zugasti, I., Torres, T. et al. (2015). Protein diagenesis in *Patella* shells: Implications for amino acid racemisation dating. *Quat. Geochronol.*, 27: 105–118.

Penkman, K.E.H., Kaufman, D.S., Maddy, D., and Collins, M.J. (2008). Closed-system behaviour of the intra-crystalline fraction of amino acids in mollusc shells. *Quat. Geochronol.*, 3: 2–25.

Penkman, K.E.H., Preece, R.C., Bridgland, D.R. et al. (2011). A chronological framework for the British Quaternary based on *Bithynia opercula*. *Nature*, 476: 446–449.

Pierini, F., Demarchi, B., Turner, J., and Penkman, K. (2016). *Pecten* as a new substrate for IcPD dating: The Quaternary raised beaches in the Gulf of Corinth, Greece. *Quat. Geochronol.*, 31: 40–52.

Radkiewicz, J.L., Zipse, H., Clarke, S., and Houk, K.N. (1996). Accelerated racemization of aspartic acid and asparagine residues via succinimide intermediates: An ab initio theoretical exploration of mechanism. *J. Am. Chem. Soc.*, 118: 9148–9155.

San Antonio, J.D., Schweitzer, M.H., Jensen, S.T. et al. (2011). Dinosaur peptides suggest mechanisms of protein survival. *PLoS One*, 6: e20381.

Schleifer, K.H. and Kandler, O. (1972). Peptidoglycan types of bacterial cell walls and their taxonomic implications. *Bacteriol. Rev.*, 36: 407–477.

Schroeder, R.A. (1975). Absence of β-alanine and γ-aminobutyric acid in cleaned foraminiferal shells: Implications for use as a chemical criterion to indicate removal of non-indigenous amino acid contaminants. *Earth Planet. Sci. Lett.*, 25: 274–278.

Schroeder, R.A. and Bada, J.L. (1976). A review of the geochemical applications of the amino acid racemization reaction. *Earth-Sci. Rev.*, 12: 347–391.

Schroeter, E.R. and Cleland, T.P. (2016). Glutamine deamidation: An indicator of antiquity, or preservational quality? *Rapid Commun. Mass Spectrom.*, 30: 251–255.

Sepetov, N.F., Krymsky, M.A., Ovchinnikov, M.V. et al. (1991). Rearrangement, racemization and decomposition of peptides in aqueous solution. *Pept. Res.*, 4: 308–313.

Simpson, J.P., Penkman, K.E.H., Demarchi, B. et al. (2016). The effects of demineralisation and sampling point variability on the measurement of glutamine deamidation in type I collagen extracted from bone. *J. Archaeol. Sci.*, 69: 29–38.

Simpson, J.P., Fascione, M., Bergström, E. et al. (2019). Ionisation bias undermines the use of MALDI for estimating peptide deamidation: Synthetic peptide studies demonstrate ESI gives more reliable response ratios. *Rapid Commun. Mass Spectrom.*, 33: 1049–1057.

Steinberg, S. and Bada, J.L. (1981). Diketopiperazine formation during investigations of amino acid racemization in dipeptides. *Science*, 213: 544–545.

Steinberg, S.M. and Bada, J.L. (1983). Peptide decomposition in the neutral pH region via the formation of diketopiperazines. *J. Org. Chem.*, 48: 2295–2298.

Sykes, G.A., Collins, M.J., and Walton, D.I. (1995). The significance of a geochemically isolated intracrystalline organic fraction within biominerals. *Org. Geochem.*, 23: 1059–1065.

Takahashi, O., Kobayashi, K., and Oda, A. (2010). Computational insight into the mechanism of serine residue racemization. *Chem. Biodivers.*, 7: 1625–1629.

Takahashi, O., Kirikoshi, R., and Manabe, N. (2017). Racemization of serine residues catalyzed by dihydrogen phosphate ion: A computational study. *Catalysts*, 7: 363.

Tomiak, P.J., Penkman, K.E.H., Hendy, E.J. et al. (2013). Testing the limitations of artificial protein degradation kinetics using known-age massive Porites coral skeletons. *Quat. Geochronol.*, 16: 87–109.

Torres, T., Ortiz, J.E., and Arribas, I. (2013). Variations in racemization/epimerization ratios and amino acid content of *Glycymeris* shells in raised marine deposits in the Mediterranean. *Quat. Geochronol.*, 16: 35–49.

Towe, K.M. and Thompson, G.R. (1972). The structure of some bivalve shell carbonates prepared by ion-beam thinning. *Calcif. Tissue Res.*, 10: 38–48.

van Doorn, N.L., Wilson, J., Hollund, H. et al. (2012). Site-specific deamidation of glutamine: A new marker of bone collagen deterioration. *Rapid Commun. Mass Spectrom.*, 26: 2319–2327.

Vallentyne, J.R. (1964). Biogeochemistry of organic matter – II Thermal reaction kinetics and transformation products of amino compounds. *Geochim. Cosmochim. Acta*, 28: 157–188.

Wallace, A.F. and Schiffbauer, J.D. (2016). Paleoproteomics: Proteins from the past. *eLife*, **5**: e20877.

Walton, D. (1998). Degradation of intracrystalline proteins and amino acids in fossil brachiopods. *Org. Geochem.*, 28: 389–410.

Wehmiller, J.F. (1977). Amino acid studies of the Del Mar, California, midden site: Apparent rate constants, ground temperature models, and chronological implications. *Earth Planet. Sci. Lett.*, 37: 184–196.

Wehmiller, J.F. (1980). Intergeneric differences in apparent racemization kinetics in mollusks and foraminifera: Implications for models of diagenetic racemization. In: *Biogeochemistry of Amino Acids* (eds. P.E. Hare, T.C. Hoering and K. King Jr.), 341–345. New York: Wiley.

Wehmiller, J. and Hare, P.E. (1971). Racemization of amino acids in marine sediments. *Science*, 173: 907–911.

Weiner, S. (1983). Mollusk shell formation: Isolation of two organic matrix proteins associated with calcite deposition in the bivalve *Mytilus californianus*. *Biochemistry*, 22: 4139–4145.

Welker, F., Collins, M.J., Thomas, J.A. et al. (2015). Ancient proteins resolve the evolutionary history of Darwin's South American ungulates. *Nature*, 522: 81–84.

Welker, F., Soressi, M.A., Roussel, M. et al. (2016). Variations in glutamine deamidation for a Châtelperronian bone assemblage as measured by peptide mass fingerprinting of collagen. *STAR: Sci. Technol. Archaeol. Res.*, 3: 15–27.

Westaway, R. (2009). Calibration of decomposition of serine to alanine in *Bithynia opercula* as a quantitative dating technique for Middle and Late Pleistocene sites in Britain. *Quat. Geochronol.*, 4: 241–259.

Wilson, J., van Doorn, N.L., and Collins, M.J. (2012). Assessing the extent of bone degradation using glutamine deamidation in collagen. *Anal. Chem.*, 84: 9041–9048.

3 Proteins in Fossil Biominerals

3.1 Bone and Other Collagen-based Hard Tissues

The analysis of biomolecules (ancient DNA, proteins) from ancient bone has been one of the major goals of research since the discovery of the survival of organic matter in fossils. This is because bone represents the direct evidence of the presence and/or activity of animals (including humans) on a site.

Bone is a highly structured biomaterial, with several hierarchies from the macro scale to the atomic level (Figure 3.1). Bone proteomes are complex, but dominated (90% of the organic mass) by collagen, which is also the protein with the highest probability of being retrieved from fossil samples (Cappellini et al., 2012; Wadsworth and Buckley, 2014). NCPs, i.e. 'non collagenous proteins', include bone morphogenetic proteins, growth/differentiation factors, osteocalcin and osteopontin, immunoglobulin, matrix extracellular phosphoglycoprotein, proteoglycans, Pro-rich proteins, lysine-rich coiled-coil proteins, ELL-associated factor, sialoproteins, ADP-ribosyl cyclase/cyclic ADP-ribose hydrolase. Although NCPs are increasingly being recovered and studied from ancient substrates, to date collagen is still the main target biomolecule.

Collagen chains are made of Pro/Hyp–Xaa–Gly repeats (i.e. proline/hydroxyproline followed by any amino acid followed by a glycine residue), twisted into a triple helix (tropocollagen), which in turn constitutes the fibrils. In bone, these fibrils are mineralized and bound to the mineral (hydroxyapatite, $Ca_{10}(PO_4)_6(OH)_2$). Many fibrils can be bound into fibres, and these form the different types of bone tissue (compact and spongy/cancellous). Neighbouring collagen molecules are offset by 67 nm, creating 'gap' regions in the fibrils (e.g. Hodge et al., 1965; Weiner and Traub, 1986).

Amino acids and Proteins in Fossil Biominerals: An Introduction for Archaeologists and Palaeontologists, First Edition. Beatrice Demarchi.
© 2020 John Wiley & Sons Ltd. Published 2020 by John Wiley & Sons Ltd.

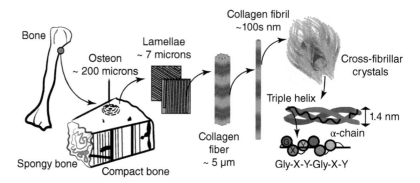

Figure 3.1 Hierarchical structure of bone, from the macroscale to the nanoscale. Image adapted from Sadat-Shojai et al. (2013) and Reznikov et al. (2018).

Recently, the fractal-like hierarchical organization of the apatite crystals and collagen has been elucidated: high-resolution transmission electron microscopy, scanning TEM (STEM) tomography imaging and 3D reconstruction, as well as electron diffraction, showed that the mineral particles are neither exclusively intrafibrillar nor extrafibrillar, but 'form a continuous cross-fibrillar phase where curved and merging crystals splay beyond the typical dimensions of a single collagen fibril' (Reznikov et al., 2018). This is very important for diagenesis studies because it gives experimental proof of the fact that bone behaves as an open system.

The open-system behaviour of bone was 'discovered' when several research groups (notably that of Jeffrey Bada in the USA) attempted to use bone as a substrate for amino acid racemization dating, especially in the 1970s. Of course, being able to date animal and human remains directly and beyond the range of radiocarbon (which is, today, around 50–60 ka BP) would have been an extremely important advance in Quaternary geochronology, enabling worldwide correlation of events occurring during the Pleistocene. A typical example of such an application was the dating of the arrival and extinction of archaic humans in the various continents. Indeed, the dating of the peopling of the Americas was one of the *causes celèbres* of AAR dating (Bada et al., 1974; Bada, 1985), which brought to the attention of scientists the fact that the diagenetic behaviour of bone proteins is extremely unpredictable.

While AAR dating of bone seems to be an impractical avenue, with recent studies highlighting that the preservation of the bone proteome and its age do *not* correlate with the extent of amino acid racemization (Wadsworth et al., 2017) and that fossil (Mesozoic) bones host thriving microbiomes (Saitta et al., 2019), the advent of proteomics by mass spectrometry has suddenly opened up new avenues of research in evolutionary biology. Being able to recover molecular sequences from extinct organisms means being able to clarify phylogenetic relationships between extinct and extant organisms (Cappellini et al., 2012, 2014; Welker et al., 2015a, 2017) that cannot, to date, be solved in any other way. A famous example is the claimed sequencing of collagen from dinosaur bones, which has fuelled more than ten years of debate in the field of molecular palaeontology (Asara et al., 2007; Buckley et al., 2008, 2017; Kaye et al., 2008; Manning et al., 2009; San Antonio et al., 2011; Cleland et al., 2015; Schroeter et al., 2017). Archaeological applications of bone proteomics, focusing on collagen but also NCPs such as fetuin or osteocalcin, include the ability to determine which lineage of ancient humans inhabited a site and created material culture (Welker et al., 2015b, 2016; Brown et al., 2016; Chen et al., 2019), the potential to detect infectious diseases, and a range of physiological and pathological information (Procopio et al., 2017; Sawafuji et al., 2017; Wadsworth and Buckley, 2018).

Antler is an example of a bone material that has great mechanical properties and that could be recovered and exploited easily in prehistory. As such, it has been widely used as a raw material for making tools or ornaments and even shamanic objects. The proteome of antler, which is true bone that grows externally on the skull of some ungulates (Cervidae, deer), is of great scientific interest because antler matures rapidly (within ~60 days) and is also shed rapidly (~5 days); antler's proteome is therefore complex, including a minimum of ~420 proteins, e.g. osteomodulin, osteopontin, osteonectin, exostosin-2, calcitonin receptor, neurofibromin and fibronectin, all related to ossification (Gao et al., 2010). Collagen is also present, allowing the successful identification of worked artefacts (e.g. combs) on the basis of peptide fingerprints (von Holstein et al., 2014). We note that horn was equally widely used as a raw material in the past, but horn has a different type of structure, typically present in bovids, which comprises a bony 'core' sheathed in a thick layer of keratin. In this respect, horn is similar to hoof, baleen, tortoiseshell and rhinoceros 'horns': all are animal hard tissues in which the main protein is not collagen but keratin; a combination of visual observations and protein analysis is usually sufficient for determining the origin of these materials (Solazzo et al., 2013, 2017; O'Connor et al., 2015).

3.2 Tooth

Tooth is another excellent source of modern and ancient proteomes (Porto et al., 2011; Castiblanco et al., 2015; Stewart et al., 2016; Parker et al., 2019). It is composed of three mineralized tissues: cementum, dentin and enamel (Figure 3.2).

Cementum binds collagen fibres which hold the tooth in place. Dentin is a bone-like phase with 20–30% organic matter by weight, which is tightly bound to the overlying layer of enamel (Bartlett, 2013). Enamel is highly mineralized and contains a mere 1% of organics, including the proteins amelogenin, ameloblastin and enamelin as well as the proteinases matrix metalloproteinase-20 and kallikrein-related peptidase-4. Enamel behaves as a closed system with respect to diagenesis (Griffin et al., 2009; Dickinson et al., 2019). Amelogenin is especially interesting because it is encoded by the single-copy amelogenin-encoding gene (AMG), which carries sequence differences between Y (male) and X (female) alleles; the PCR amplification of this gene or the direct sequencing of the peptides by mass spectrometry are therefore useful methods for sexing skeletons of immature or incomplete individuals in the archaeological record (Faerman et al., 1995; Stone et al., 1996; Stewart et al., 2016; Parker et al., 2019). It is likely that enamel will be one of the main substrates for ancient protein studies in the future, as it has shown that proteins are effectively occluded in the inorganic matrices and can survive into deep time, allowing phylogenetic inferences for a genus of rhinoceros from the 1.8 Ma site of Dmanisi (Cappellini et al., 2019; Dickinson et al., 2019). The next step will be to apply similar approaches to hominin remains from this and other sites.

Dentin is a good source of DNA and proteins, and while its open-system behaviour makes it a problematic substrate for AAR dating, it has been successfully used in forensic applications, i.e. for estimating the age at death of unidentified bodies (Ogino et al., 1985; Masters, 1986; Ritz et al., 1990; Ohtani, 1994; Ohtani and Yamamoto, 2010). Similar approaches have also been developed for enamel (Griffin et al., 2009). Of note, dental plaque or calculus is also a wonderful repository of ancient biomolecules, particularly those related to diet and disease (Warinner et al., 2014a) and proteomics applications have already begun to rewrite the history of human and environment co-evolution, for example with regard to the origin of dairy practices (Warinner et al., 2014b; Jeong et al., 2018).

Finally, ivory (Figure 3.2) is derived from the tusks and teeth of elephants and mammoths, walruses, sperm whales, killer whales, narwhals, hippopotamuses and warthogs. Composed mainly of dentin, ivory has been used for millennia as a highly prized material for art objects. The high collagen content of dentin (Jágr et al., 2012) makes this material ideal for ancient protein studies, which are able to identify the origin of the material (Coutu et al., 2016).

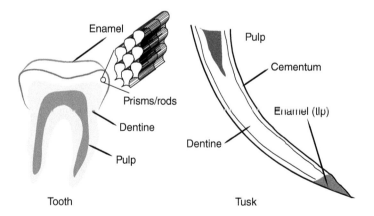

Figure 3.2 Structure of human tooth and of elephant tusk. Adapted from: www.smartwaterfoundation.org/endangered-species

3.3   Eggshell

Eggshell is one of the most useful but underused resources in archaeology and palaeontology: it has obvious symbolic significance (let us just think about the innumerable world creation myths, or religious rituals, which centre around the egg as a symbol of creation and rebirth) and its presence in an archaeological site testifies directly to the choices made by humans in selecting birds for consumption or breeding. At the same time, it gives us an immediate snapshot of the environmental conditions at the site (for example, a large assemblage of water birds indicates that a body of water would probably have been present close by). Furthermore, eggshell survives well in alkaline and neutral soils (conditions that are widespread worldwide) and as such it acts as a protective capsule for the proteins embedded in the mineral. The eggshell proteome has become an increasingly attractive model system for biomineralization, because it is the fastest-forming calcified tissue in nature (Wu et al., 1995; Rodríguez-Navarro et al., 2015). Studies on modern and ancient eggshell have revealed that the proteome trapped in the calcite crystals is highly complex, with hundreds of proteins being recovered (Mann, 2015). This is largely due to the way in which the egg and its shell are formed within the bird's womb.

The egg is formed in the ovary and transferred to the oviduct, where the yolk resides for times of varying lengths: fertilization occurs in the infundibulum if sperm is present (15 minutes), after which the albumen is secreted around the yolk (magnum, 3 hours) and the membranes, inner and outer, are formed (isthmus, 1 hour). Mineralization occurs in the uterus, where highly crystalline calcite (composed of large columnar crystals, the prisms) is produced in about 20 hours (in hens). In the uterus, the forming egg is immersed in an acellular fluid that is packed with both the organic and inorganic precursors necessary for the formation of the shell. These include calcium and carbonate ions, which are transported actively and released into the uterine fluid from the bloodstream, thanks to ionic transporters, and which are hydrated thanks to carbonic anhydrases (Jonchère et al., 2012). Other organics are proteins, polysaccharides and proteoglycans, which promote and regulate mineralization.

The details of these mechanisms at the organic–inorganic interface are unknown, but it has recently been discovered that calcite formation begins with the accumulation of amorphous calcium carbonate (ACC) on the eggshell membranes, specifically on the so-called mammillary cores (Rodríguez-Navarro et al., 2015) (Figure 3.3). ACC acts as a transient storage of calcium carbonate which can be easily mobilized, and its importance has been demonstrated for several organisms. The flat disk-shaped ACC undergoes direct transformation to calcite, which then grows the large crystals of the palisade layer.

This experimental evidence is in agreement with the hypothetical mechanism of calcite formation based on a molecular dynamic study

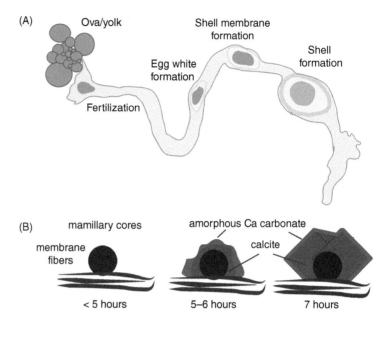

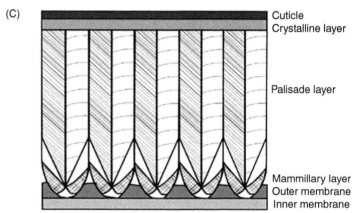

Figure 3.3 (A) Schematic representation of the process of egg production. (B) Hypothesis of the mechanism of fast calcite mineralization in eggshell: after around 5 hours the organic mammillary cores on the membranes are covered by ACC, which aggregates from spherical nanoparticles. In the following 2 hours, secondary nucleation transforms the ACC into calcite crystals and the growth process continues until only large crystals (surrounded by a film of AAC) remain. Adapted from Rodríguez-Navarro et al. (2015). (C) General microstructure of the eggshell, showing the mammillary and palisade (large calcite crystals) layers, and the outer crystalline layer, covered by the cuticle.

(Freeman et al., 2011) of the molecule of ovocleidin-17 (Mann and Siedler, 1999). This protein was shown by Freeman and colleagues to be capable of transforming ACC into calcite via a catalytic cycle. Furthermore, a recent study has proposed that EGF-like repeats and discoidin-like domains 3

(EDIL3) and, to a lesser extent, milk fat globule-EGF factor 8 (MFGE8), two glycoproteins that are consistently detected in eggshell proteomes, guide the ACC-containing vesicles to the mineralization site (Stapane et al., 2019).

The fact that all bird eggshells are made entirely of calcite suggests that the mechanism of eggshell formation ought to be fairly universal, and for a while ovocleidin-17 and other highly structured calcium-binding proteins called C-type lectins, for example struthiocalcin, dromaiocalcin and rhea-calcin isolated from ostrich, emu and rhea eggshell, respectively (Mann, 2004; Mann and Siedler, 2004) were thought to be universal in eggshell. Further, these proteins have been shown to be dominant in the intracrystalline fraction of eggshell (e.g. in ostrich eggshell). Therefore, one should expect to find C-type lectins consistently in eggshells from all avian species.

One way to test this is to analyse the intracrystalline proteome isolated from eggshell of different avian taxa. As a result of the rapid process of eggshell growth, the organic phase (uterine fluid) is occluded in the crystals, and, as such, the whole uterine proteome can be trapped and preserved for hundreds, thousands or even millions of years. This is key, because it allows the use of museum collections of eggshells in order to characterize the proteins, rather than having to rely on fresh specimens, which are difficult to obtain thanks to the legislation protecting nests and eggs of wild birds. Importantly, the isolation of the intracrystalline fraction by bleaching does not drastically reduce the number or the spectrum of the proteins trapped in the shell (e.g. 273 were identified in bleached ostrich eggshell: Demarchi et al., 2016) but bleaching does remove any exogenous contamination. Therefore, the composition of the bleached fraction is a representative portion of the whole eggshell proteome.

A comparative proteomics study of 23 species of bird eggshell (Presslee et al., 2018), conducted using eggshell specimens from museums or available commercially, revealed that in fact C-type lectins are not universal components of the bleached eggshell proteome. Instead, the most frequently occurring proteins were:

- von Willebrand Factor D domain superfamily, which includes IgGFc-binding protein, mucin 5-AC and mucin 5-B, as well as ovomucin – mucins are gel-forming proteins that have been linked to biomineralization in mollusc shells (Marin et al., 2000) and have been identified in eggshell membranes, most recently in a proteomics study which identified a 1–10 fold increase of mucin 5AC in fertilized egg membranes (Cordeiro and Hincke, 2016);
- the 'BPI-fold' family, i.e. bactericidal permeability-increasing proteins which include ovocalyxin-36 (OCX-36), isolated and characterized in chicken eggshell (Gautron et al., 2011);
- the serine protease inhibitors (SERPIN family): ovalbumin, alpha-1-antiproteinase-2, alpha-1-antitrypsin-like, alpha-2-antiplasmin, antithrombin-III, neuroserpin, pigment epithelium-derived factor, plasma protease C1 inhibitor, serine protease inhibitor 2.1-like, serpin B11 and serpin B4;

- ovotransferrin, melanotransferrin, iron-binding protein and sero-transferrin, which are involved in the uptake of a variety of soluble substrates such as phosphate, sulfate, polysaccharides, lysine/arginine/ornithine and histidine;
- the immunoglobulin superfamily, which includes immunoglobulin, neuroglia, cell surface glycoproteins and membrane glycoproteins, such as butyrophilin and chondroitin sulfate proteoglycan core protein: the general function of immunoglobulins are antigen binding, fixation of complement, binding to various cells, e.g. lymphocytes;
- albumins, i.e. proteins able to bind cations, fatty acids and bilirubin;
- ovocleidin-116-like, which is the eggshell ortholog of mammalian matrix extracellular phosphoglycoprotein (MEPE) (Bardet et al. 2010; Mann and Mann 2013; Hincke et al. 1999) – MEPE are components of the extracellular matrix of bone and dentin and regulate bone mineralization;
- C-type lectins are mineralization-specific proteins found in a variety of bird species, notably ratites, birds of prey and chicken; while chicken eggshell seems to contain only one C-type lectin (OC-17), ostrich and other ratites have two slightly different forms (struthiocalcin-1 and -2, dromaiocalcin-1 and -2, rheacalcin-1 and -2); further, OC-17 carries a net positive charge and it has been shown to interact with the mineral matrix via Arg or Lys (basic) residues, while the ratite-specific forms carry a negative charge; a homologue of OC-17 was isolated from goose eggshell – ansocalcin (Lakshminarayanan et al., 2003) – but its sequence has now been removed from NCBI.

Eggshell has been one of the most successful substrates for ancient protein studies: AAR analyses of ostrich eggshell from African archaeological sites have revolutionized our understanding of the importance of closed-system minerals for geochronology, showing that amino acids enclosed tightly within a mineral matrix represent the degradation products of proteins degraded in situ and can therefore yield reliable information on the time elapsed since the deposition of the (fossil) egg. Brooks et al. (1990) were the first to test the potential of eggshell for AAR dating in a major study, but the work of Giff Miller on extinct and extant ratite eggshells (Miller et al., 1992; Johnson et al., 1997, 1998) has been especially important because, for the first time, it combined protein diagenesis for geochronology and palaeothermometry with palaeoenvironmental reconstructions on the basis of the geochemical signals registered in the stable isotope composition of the eggshell carbonate. Ancient amino acids trapped in emu eggshell from the Quaternary deposits of Australia have proven very useful for reconstructing temperature changes (cooling) in the Southern Hemisphere during the Last Glacial Maximum; this was achieved by exploiting the closed-system behaviour of eggshell, because the extent of racemization in such systems is related to time and temperature only (without other confounding factors, such as pH, water table fluctuations, etc.). Another excellent example of the potential of eggshell for palaeoenvironmental studies is the work conducted by Johnson and colleagues at Equus Cave in South Africa (Johnson et al., 1997). Here, radiocarbon-calibrated AAR ages were used to establish a stratigraphic sequence of ostrich eggshell fragments spanning the last 17 000 years. The same ostrich eggshells were analysed for their stable isotope composition (oxygen, nitrogen and carbon), from which it could be determined that, between 17 and 12 ka BP, the palaeovegetation had been dominated by C_3 plants and that there was decreased rainfall, reduced evaporation and lower temperatures compared to today (Johnson et al., 1997). More recently, protein degradation patterns typical of burning have been detected in *Genyornis* eggshells (Miller et al., 2016); as a consequence, the main cause of extinction of these giant birds was deemed to be human predation. With the advent of more sophisticated methods of analysing protein sequences and other biomolecules (e.g. DNA), eggshell has become one of the most promising substrates for the nascent field of biomolecular archaeology.

3.4 Mollusc Shell

Mollusc shell is one of the most common materials found in archaeological and geological sites. Shells are widespread in the fossil record because theirs is a highly populated phylum, with more than 80 000 extant species (and many fossil ones). Molluscs have adapted to living in seawater (from the intertidal zone to the great depths of the oceans), and freshwater (from fast-flowing rivers to small ponds), but also on land (including deserts and high-altitude mountains).

Molluscs were exploited by humans as a food resource throughout prehistory; as the empty shells were discarded over time, they accumulated to form 'middens', which can be up to several meters in height. Indeed, the planning and foresight necessary for effective shell gathering on the shoreline, especially the intertidal zone, is one of the arguments in favour of an early appearance of 'modern' behaviour in humans. For example, in South Africa, the earliest evidence for shellfish collection has been dated to 164 ka BP, from the deposits at Pinnacle Point Cave 13B (Marean et al., 2007). The analysis of the malacofauna at Pinnacle Point and comparisons with ethnographic studies suggest that, for a successful shellfish gathering, it was necessary to not only calculate tidal movements and accessibility to the shore, but also to cooperate with others (Jerardino and Marean, 2010). As a result, exploitation of coastal resources in general, and molluscs in particular, has in the last decade become a new paradigm in prehistoric archaeology to signal 'advanced' intellectual abilities of archaic and modern humans alike. For example, the shell middens and remains of marine mammals found within Neanderthal archaeological layers at Gorham's and Vanguard caves, Gibraltar, have been central to the debate on the uniqueness of modern traits in *Homo sapiens* (Stringer et al., 2008).

Mollusc shells have excellent material qualities and are aesthetically pleasing. It is therefore no surprise that they were also selected as raw materials for making personal ornaments as well as functional tools. The earliest material evidence of this use comes from a handful of Middle Stone Age / Aterian sites in South Africa (Blombos Cave), North Africa (Grotte des Pigeons in Morocco, Oued Djebbana in Algeria) and the Levant (Qafzeh, Skhul) between 90 000 and 70 000 years ago: perforated shells (*Glycymeris*, *Nassarius*) are found together with engraved ostrich eggs and ochre and refined stone tools (d'Errico et al., 2005; Bouzouggar et al., 2007; Bar-Yosef Mayer et al., 2009). Shell is thus one of the first materials to be transformed into 'cultural heritage' objects. As such, being able to access the molecular information locked within a shell's mineral skeleton is especially important (Sakalauskaite et al., 2019).

Finally, mollusc shells offer direct evidence of past climatic changes: the sea level rose and fell several times in the past, due to the succession of glacial–interglacial cycles and, by doing so, dramatically transformed the coastline. The continental shelf emerged during low sea-level periods, and beaches were formed which marked the new coastline. Conversely, these coasts were eroded and submerged during highstands, when the sea rose to >100 m above the present datum (Chappell and Shackleton, 1986). A classic illustration of this effect is the emergence of land bridges during lowstands, such as the one between the British Isles and continental Europe, or the Bering Land Bridge uniting Alaska and Asia. As many highstands reached or exceeded the present sea level, many fossil shorelines can today be found above present sea level, appearing to have been 'raised'. The molluscan fauna embedded in these fossil beaches can be an excellent aid for reconstructing palaeoenvironments (on the basis of their ecological requirements and their stable isotope composition). Furthermore, they are a very good substrate for AAR dating, which can thus be used to provide chronological information on the age of these fossil beaches (and by inference on glacial–interglacial cycles).

The most striking feature of most mollusc shells (especially those of Bivalvia and Gastropoda) is their complex microstructure (Figure 3.4), which includes an organic outer layer (the periostracum) and several calcified inner layers. The latter are often made of large elongated crystals (of calcite or aragonite) or small hexagonal tablets of aragonite (the nacreous layer or 'mother-of-pearl'). However, the variety of crystal morphology, size and mineral composition is staggering (see the comprehensive classification by Carter, 1989), so much so that they can represent a way of identifying the shell taxon. The exceptional material properties of molluscs derive from the juxtaposition of different microstructural layers (some are harder but quite fragile, others are easily fractured but quite flexible) and mineralogies, as well as the fact that introducing an interface between different layers makes the propagation of fracture more difficult (thus enhancing the protection of the animal).

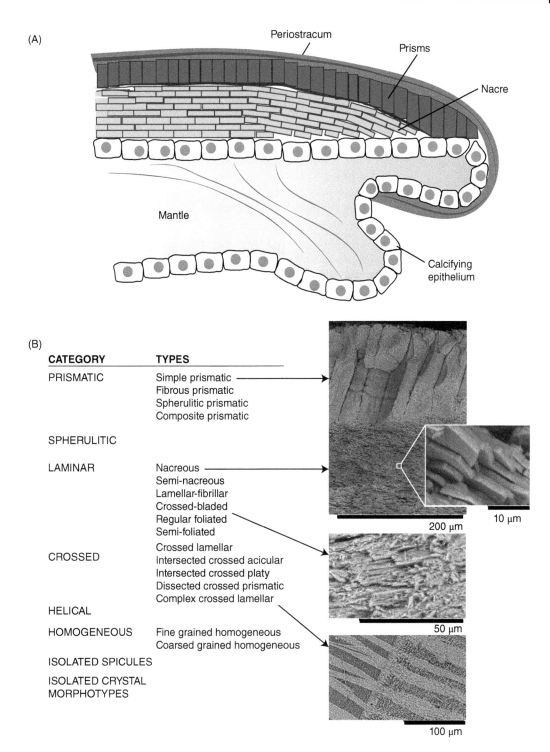

Figure 3.4 (A) Schematic representation of the calcifying epithelium and its localization within the mollusc shell. (B) Summary of the main shell microstructures according to Carter (1989) and summarized in Marin et al. (2012).

Shells, especially those with mother-of-pearl layers, have been one of the most studied model organisms for biomineralization, due to the importance of nacre both commercially and as a template for bioinspired materials with exceptional mechanical properties. Despite a long history of research, the formation of the shell is still debated. It is known that the shell is deposited due to the action of the mantle, a biological feature that covers the inner surface of the shell (Figure 3.4). Different zones of the mantle are dedicated to the secretion of the molecular machinery ('mantle secretome') in charge of the formation of different shell microstructures, as demonstrated by studies on the expression of different genes (Kocot et al., 2016). Within a specialized area of the mantle, a sheet of protein (the periostracum) is secreted, representing the substratum on which the outer crystalline shell layer is deposited. Between the mantle and the inner shell surface (called the extrapallial space), the extrapallial fluid is the medium in which both the organic matrix and the crystalline components are formed. In this aqueous microenvironment, inorganic ions (including Na^+, K^+, Ca^{2+}, Mg^{2+}, Cl^-, SO_4^{2-}, CO_3^{2-}, and HCO_3^-) are transported from the external environment and achieve the appropriate supersaturation of $CaCO_3$ necessary for shell crystal growth. Various organic macromolecules, including proteins, polysaccharides and lipids, are secreted by mantle cells or transported from elsewhere to the extrapallial fluid 3.4. According to the scheme put forward by Addadi et al. (2006) and summarized in Marin et al. (2012), the first step for shell formation is the secretion of the organic framework, made of chitin sheets sandwiched between proteins (the interlamellar matrix). Between sheets, amorphous mineral precursors are assembled and nanocrystals nucleated thanks to polyanionic molecules. The nanocrystals are then self-assembled to form a meso-crystal – this concept was put forward by Cölfen and Antonietti (2005). The fusion of the mesocrystals part occludes some macromolecules, the latter becoming trapped within the fully formed crystal (and representing the intracrystalline matrix).

The biomolecules involved in shell mineralization (or at least trapped within the crystals) belong to different categories (Marin et al., 2012): polysaccharides (chitin and acidic sugars, the latter forming proteoglycans by covalently binding proteins), lipids (fatty acids, cholesterol, triglycerides, ceramides), pigments (carotenoid and polyenes), small peptides and free amino acids (which are consistently found in the intracrystalline matrix of shells in all AAR studies), and proteins. Overall, the organic component represents 0.01–5 wt% of the shell, and the protein fraction is the dominant one.

The characterization of the proteome of the shell ('shellome') has been one of the most fruitful and exciting lines of research in biomineralization for the past 20 years. The first approaches involved the characterization of the amino acid composition (which led to the somewhat over-simplistic distinction between acidic/soluble/intracrystalline and hydrophobic/insoluble/intercrystalline matrices), and then of the mass and pI of the molecule (using SDS-PAGE). These were then complemented by the direct sequencing of proteins extracted from the mineral matrix; this was the method of choice between the late 1990s and the early 2000s, and resulted in around 40 to 50 protein sequences being deposited in public databases, mainly from the edible abalone, *Haliotis*, and the pearl oyster, *Pinctada*. The sense of the complexity of the shellome, however, increased dramatically with the first transcriptome-based studies, which revealed that hundreds of proteins are encoded in the mantle secretome, and that the majority of these have a completely unknown function (Jackson et al., 2010)! Proteomics approaches on the other hand were able to determine de novo the primary sequence of several peptides from a variety of model organisms, but, in the absence of reference genomes or transcriptomes, these peptides were left 'floating'. This exercise, however, revealed an interesting fact: shell proteins are 'modular', i.e. some functional blocks are swapped around in different proteins (Marin et al., 2012). The most successful approaches so far have been those in which the peptidome is reconstructed de novo (using soft-ionization tandem mass spectrometry and bioinformatics) and each peptide sequence is searched against a database of transcripts (ESTs databases) for that organism (Joubert et al., 2010; Marie et al., 2010; Berland et al., 2011; Mann and Edsinger, 2014). This allows the filtering out of ESTs that are not involved in biomineralization and, with respect to protein sequences, allows clarification of some of their functions and of whether they have deep origins or not (Marie et al., 2017).

The most intriguing discovery resulting from this approach is that, once again, molecular evolution and morphological differences do not coincide. A study by Marie et al. (2017) focusing on nacre-associated proteins from two Unionida shells and other model organisms (Figure 3.5) shows that some of the molecular machinery is in fact deeply conserved and could well represent a fraction of an original ancestral biomolecular toolkit. However, the analysis of mantle secretomes in different species shows that the proteins secreted by the mantle are markedly different, even more than would be expected on the basis of the external appearance of the shell (Kocot et al., 2016). This implies either independent evolution of the mollusc species or extremely rapid evolution of the mantle secretome. Either way, it is clear that the set of genes responsible for shell formation is large and diverse, and not limited to a subset of conserved developmental genes (McDougall and Degnan, 2018). In fact, it seems that the gene regulatory network is constituted by a core, which coordinates gene expression, controlling timing and expression of downstream effector genes, while the terminal branches of the network include genes encoding proteins (secreted by the mantle), lipids and polysaccharides (McDougall and Degnan, 2018). Overall, the likely mechanisms by which the secretome evolved are the following (Kocot et al., 2016; Aguilera et al., 2017; McDougall and Degnan, 2018):

- co-option of highly conserved genes (such as those encoding core shared domains, e.g. carbonic anhydrases, protease inhibitors, peroxidases, alkaline phosphatases tyrosinases, chitin-binding 2, and von Willebrand factor-A), which probably occurred several times independently in different molluscan lineages;
- rapid evolution of gene families, which have resulted in paralogues that sometimes are species-specific, e.g. lysine (K)-rich mantle proteins (KRMPs);

"Shellome" composition (from Marie et al., 2017)		Nacre				Prisms		Prisms and foliated calcite	Prisms & cross-lamellar	
		Elliptio complanata	Villosa leinosa	Pinctada spp.	Mytilus spp.	Pinctada spp.	Mytilus spp.	Crassostrea gigas	Lottia gigantea	Caepea nemoralis
RLCD - containing	Ala-rich									
	Gly-rich									
	Met-rich									
	Ser-rich									
	Gln-rich									
	Phe-rich									
	Pro-rich									
	Tyr-rich									
	Thr-rich									
	Val-rich									
Extracellular matrix	VWA									
	VWC									
	SCP									
	CCP/SUSHI									
	EGF									
	ZP									
	IGF-BP									
	FN3									
	Cyclophylin									
	ANF_receptor									
	FAM20C									
	MSP-like									
	Filament-like									
Enzyme	Peroxidase									
	Tyrosinase									
	Copamox									
	CA									
	SOD									
	TRX									
	Apase									
Ca-binding	Asp-rich									
	Glu-rich									
	EF-hand									
	GN-repeat									
	Calponin									
Polysaccharide interacting	Glyco_hydro_18									
	Glyco_hydro_20									
	Glyco_hydro_23									
	Glyco_hydro_31									
	CBD2									
	Chit_bid_3									
	LamG									
	CLECT									
Immunity	WAP									
	Kunitz-like									
	TIMP									
	Peptidase C1									
	macroglobulin									
	Peptidase S1									
	C1q									
	Lysosyme-G									
	IgG-type2									
	Lactamase									
Other orphans		8	12	5	11	7	14	18	14	12

Figure 3.5 Summary of protein families found in the shellome extracted from different mollusc species and microstructural types (from Marie et al., 2017).

Apart from core shared domains, shell biomineralization also seems to be characterized by other underlying principles. For example, a high proportion of proteins contain regions that are of low complexity (and with a repetitive composition biased towards Ala and Gly). Importantly, these repeated low complexity domains (RLCDs) are easily swapped, added or lost via unequal crossover and replication slippage, and are likely contributors towards the rapid evolution of the secretomes.

It is evident that, while phylum Mollusca is one of the most interesting groups of biominerals with regard to the amount of palaeoecological and archaeological information preserved in their chemical and biomolecular composition, it is perhaps also the most complex. In practice, this means studying each species and each microstructural layer as a semi-independent feature and then comparing their molecular makeup in order to draw some general conclusions. The complexity of the shellome in particular is a complicating factor, with respect to both the application of palaeoproteomics to taxonomic identification (which will be explored in Chapters 4 and 5), as well as, to a certain extent, the 'upgrading' of protein diagenesis geochronology, i.e. from a relative method which yields D/L values that can be calibrated, to a numerical method in its own right.

3.5 Other Substrates

In order to conclude the overview of biomineralized proteomes which may be of interest to archaeologists and earth scientists, we will briefly discuss a few other groups of animals that produce biomineralized skeletons and that are found in archaeological and geological deposits:

- coral
- foraminifera
- brachiopods
- arthropods

This fauna is important because it may represent an accurate proxy of the palaeoenvironmental conditions at the site, always assuming that the ecological requirements of each animal were the same in the past. In some instances these organisms were also exploited directly by humans, either as a subsistence resource or as raw materials.

3.5.1 Coral

Coral (Anthozoa, Cnidaria) is perhaps one of the most studied organisms in biomineralization and environmental sciences: its mineralization is biologically induced and controlled and is very sensitive to environmental change. The 'bleaching' phenomenon, that is the death of corals that make up the Great Barrier Reef, for example, as a direct consequence of global climate change and ocean acidification is well known and an unfortunate reality. Coral has also been an important prestige raw material in prehistory, since at least the Neolithic, although it is typically associated with Celtic ornaments and the Iron Age (Skeates, 1993; Morel et al., 2000; Borrello, 2001; Fürst et al., 2016). However, the use of the terminology 'coral' in prehistory might be misleading: Cnidaria are a hugely varied group from the morphological point of view. The coral used for prehistoric ornaments was in fact a species of octocoral, *Corallium rubrum* (Mediterranean coral), which lives in dark areas (caves, cliffs) between a few meters to several hundred meters in depth (Taviani, 1997). Its procurement is difficult, which explains its rarity in the archaeological record. However, the difficulty in identifying *C. rubrum*, especially when diagenetically altered and without the characteristic red coloration, is another reason for this relative underrepresentation.

The body units of corals are polyps (sac-like animals, only a few millimeters in dimension, Figure 3.6), thousands of genetically identical versions of which form a colony, which in turn grows on a mineral skeleton. Although the organic matrix associated with this mineral skeleton is dominated by lipids and sugars, it is the proteinaceous component that has been studied the most (Falini et al., 2015).

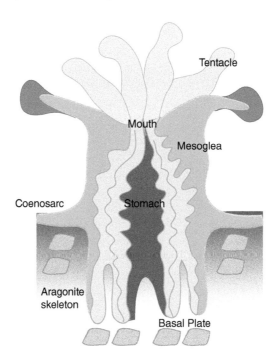

Figure 3.6 Schematic representation of a coral polyp. Source: https://commons.wikimedia.org/wiki/File:Coral_polyp_de.svg.

Coral-matrix proteins extracted from the staghorn coral *Acropora* include acidic proteins, extracellular matrix proteins, enzymes, proteins with transmembrane domains and 'orphan' proteins, i.e. proteins that cannot be simply affiliated to a specific group or function (Ramos-Silva et al., 2014).

An important finding of coral proteome-wide studies (Drake et al., 2013; Ramos-Silva et al., 2013, 2014) is that many of the identified proteins are in fact transmembrane proteins, which can be cleaved by peptidases, and that these peptidases have also been found to be an important component of the matrix. Furthermore, all the identified peptides correspond to the portion of the protein which lies outside the membrane. Therefore this hints at a physical mechanism whereby the transmembrane proteins are cleaved and the external parts are occluded in the growing skeleton – and thus should be found in sub-fossil corals. However, to date, only a handful of studies have focused on the characterization of the proteins extracted from subfossil corals, and these were essentially aimed at testing the suitability of coral as a substrate for AAR geochronology (Goodfriend et al., 1992; Hendy et al., 2012; Tomiak et al., 2013). The use of AAR on coral is particularly desirable because corals are well suited for uranium-series dating (Rink and Thompson, 2015). In general, corals are another excellent substrate for ancient protein studies and the development of proteomics techniques will certainly enhance their role in reconstructing human–environment interactions in the past.

3.5.2   Foraminifera

Foraminifera are unicellular organisms (amoeboid protozoans), which produce a mineral test (or shell). The test is usually made of calcium carbonate. Foraminifera are present worldwide in all marine and estuarine environments (and some also in freshwater or soil). At the time of writing, the World Foraminifera Database counts 8931 extant species and 40 379 fossil species (http://www.marinespecies.org/foraminifera/)! Dead foraminifera cover the ocean floor and have been used for several decades in order to reconstruct past climatic shifts: the stable isotope composition of the shells reflects water chemistry at the time of shell formation. Stable oxygen isotope ratios measured on the shell can be used to infer past water temperatures (Emiliani, 1954, 1955), but almost every trace element and stable (or radiogenic) isotope can be used for tracking past changes in seawater chemistry and biogeochemical cycles (Kucera, 2007). With regard to the relevance of foraminifera for archaeology, they have been used mainly for palaeoenvironmental reconstructions, especially in coastal settings, but also for tracking the provenance of certain stones (marble, chalk) used to make buildings or artefacts (Jones, 2014). Nummulites (a type of large foraminifera, Figure 3.7) have also been used as a raw material for ornaments, as recently discovered thanks to chemical and physical analyses carried out on Iron Age fibulae which were previously thought to be made of coral (see a summary in Fürst et al. 2016).

Figure 3.7 Nummulite: diameters can vary between 1.5 and 5 cm. Source: https://commons.wikimedia.org/wiki/File:NummuliteLyd.jpg.

Foraminifera mineralization is biologically induced and biologically controlled, but almost nothing is known about the proteome composition of these organisms. Early studies focused on the extraction of chiral amino acids and the quantification of the extent of racemization from oceanic calcareous forams for dating purposes and for palaeothermometry (Wehmiller and Hare, 1971; Bada and Schroeder, 1972, 1975; Wehmiller, 1980). Also, because the deep ocean offers the important advantage of stable temperatures (i.e. no seasonal/daily variation), the kinetics of racemization can be modelled more easily. Some of the most successful studies of racemization kinetics, carried out using both heating experiments and fossils, have in fact been conducted on foraminifera tests (Müller, 1984; Kaufman, 2006).

The amino acid composition of foraminifera was also used in one of the first chemotaxonomic studies by Haugen et al. (1989). The authors found that four foraminifera taxa could be separated out and distinguished to the level of genus on a principal component analysis plot. This early study was very promising, because the phylogenetic relationships and classification of foraminifera is based entirely on their shell, and so a molecular tool able to complement this would be extremely useful. Indeed, the first analyses aiming at the characterization of the proteinaceous matrix extracted from modern foraminifera followed not long after this study, and highlighted the fact that the mineralizing matrix was difficult to isolate. The amino acid composition was typical of an acidic protein family and a hydrophobic-rich protein family (although this earlier distinction in classes has been superseded, at least for mollusc shells, as we have seen earlier) (Robbins and Donachy, 1991). 2D gels on different extracted fractions showed the presence of proteins with molecular weights between 15 and 110 kDa, but no attempt at sequencing these molecules was carried out at the time (Stathoplos and Tuross, 1991). Later studies revealed that the mineral-associated proteins were enriched in Asx, and that DNA and actin could be recovered (Stathoplos and Tuross, 1994). This Asx enrichment was confirmed by immunological analyses and partial sequencing of a foram-associated protein, which was found to have a poly-Asp domain at the N terminus (Robbins et al., 1993). Until 2014 no proteomics studies were carried out using tandem mass spectrometry: the first study, by Sabbatini et al. (2014), concerned the matrix of the benthic foraminifera *Schlumbergerella floresiana*. They were able to reconstruct peptide sequences using a de novo approach and to identify sequences similar to collagen or with calcium-binding properties. However, when they tried to match these to known protein sequences from other biomineralized organisms, they obtained very little correspondence. In fact, this result once again confirms that the evolution of biomineralized organisms did not follow a simple pathway.

3.5.3   Brachiopods

Brachiopods or 'lamp shells' are marine invertebrates that look at first sight like molluscs, but lack the symmetry of the latter (Figure 3.8). In fact, they are composed of two valves, a ventral and a dorsal (as opposed to the left/right symmetry of molluscs). They are extremely abundant in the fossil record, and are one of the first model organisms used for biomineralization studies. Jope (1967, 1977) pioneered work on the biochemistry of the organic matrix from Brachiopoda. The fact that some of the shells are calcareous and others phosphatic was an added bonus for these early biomineralization studies. Jope identified Gly and Ala (but also Pro) as major components of the matrix, and pointed out that the amino acid composition retains a taxonomic signal. Further studies demonstrated that some brachiopod calcification proteins can be preserved in the fossil record (Endo et al., 1995). While no palaeoproteomics studies have been successfully carried out so far, brachiopods have been investigated from the point of view of modern proteome characterization. Isowa et al. (2015) identified 40 matrix proteins in *Laqueus rubellus*, of which 35 did not have a corresponding homologue in the databases. Proteins with a database homologue included MSP-130, a skeletal protein identified from sea urchins, extracellular copper/zinc superoxide dismutase, actin I, cathepsin L-cysteine proteinase, and ICP-1 (intra crystalline protein-1), which is a shell matrix protein extract from the calcitic shell of three brachiopod species (*Neothyris lenticularis*, *Calloria inconspicua* and *Terebratella sanguinea*). Similarly, Jackson et al. (2015) analysed the sequence similarity of 66 proteins from *Magellania venosa* with a range of model organisms and found a low degree of homology, and Immel et al. (2015) highlighted the extreme paucity of homologues found in the proteomes of three species of Terebratulida. Once the proteome composition of

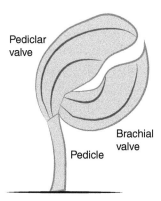

Figure 3.8 Schematic representation of a brachiopod.

modern brachiopods is better understood, the study of fossil molecules will be especially interesting, due to the abundance of fossil species of this class and their poorly understood phylogeny (Carlson, 2016).

3.5.4 Arthropods

Class Arthropoda is very diverse, including invertebrates with jointed exo-skeletons, crustaceans, insects and ostracods. These are all excellent indicators of palaeoenvironmental conditions and of human exploitation strategies. Beetles (Coleoptera) have a long history as palaeotemperature proxies (Lowe and Walker, 2014), while the marginal yet consistent exploitation of crabs, goose barnacles and sea urchins has recently been reassessed by Gutiérrez-Zugasti (2011) for the shell midden record of the Atlantic coast of Spain.

The chitinous skeletons can survive in the fossil record, although taphonomy experiments have shown that the composition of the degraded exoskeleton is very different from the living one, including an important aliphatic hydrocarbon component, and an aromatic fraction at times. Both derive from the incorporation of fatty acids in the chitin matrix during early

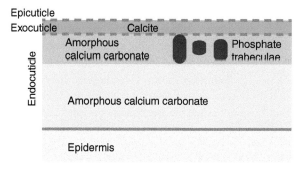

Figure 3.9 Microstructure of lobster cuticle. Simplified drawing adapted from Kunkel and Jercinovic (2013).

diagenesis (Gupta and Summons, 2011). In living crustacean groups, parts of the skeleton/cuticle are mineralized by precipitation of crystalline (or, to a lesser extent, amorphous) calcium carbonate into the twisted lamellar structure of the chitin–protein matrix (Romano et al., 2007) (Figure 3.9).

3.5.5   Conclusions

In summary, our knowledge of biomineralized proteomes is only beginning to shed some light on the evolutionary patterns of these complex and fascinating organisms. Biominerals are increasingly recovered from the archaeological and geological record and, as a consequence, we can look forward to adding information gathered from extinct biominerals to the picture, and to providing a better understanding on why and how biomineralization came to happen.

References

Addadi, L., Joester, D., Nudelman, F., and Weiner, S. (2006). Mollusk shell formation: A source of new concepts for understanding biomineralization processes. *Chemistry*, *12*: 980–987.

Aguilera, F., McDougall, C., and Degnan, B.M. (2017). Co-option and de novo gene evolution underlie molluscan shell diversity. *Mol. Biol. Evol.*, *34*: 779–792.

Asara, J.M., Schweitzer, M.H., Freimark, L.M. et al. (2007). Protein sequences from mastodon and *Tyrannosaurus rex* revealed by mass spectrometry. *Science*, *316*: 280–285.

Bada, J.L. (1985). Aspartic acid racemization ages of California Paleoindian skeletons. *Am. Antiq.*, *50*: 645–647.

Bada, J.L. and Schroeder, R.A. (1972). Racemization of isoleucine in calcareous marine sediments: Kinetics and mechanism. *Earth Planet. Sci. Lett.*, *15*: 1–11.

Bada, J.L. and Schroeder, R.A. (1975). Amino acid racemization reactions and their geochemical implications. *Naturwissenschaften*, *62*: 71–79.

Bada, J.L., Schroeder, R.A., and Carter, G.F. (1974). New evidence for the antiquity of man in North America deduced from aspartic acid racemization. *Science*, *184*: 791–793.

Bardet, C., Vincent, C., Lajarille, M.-C. et al. (2010). OC-116, the chicken ortholog of mammalian MEPE found in eggshell, is also expressed in bone cells. *J. Exp. Zool. B Mol. Dev. Evol.*, *314*: 653–662.

Bartlett, J.D. (2013). Dental enamel development: Proteinases and their enamel matrix substrates. *ISRN Dent.*, *2013*, 684607.

Bar-Yosef Mayer, D.E., Vandermeersch, B., and Bar-Yosef, O. (2009). Shells and ochre in Middle Paleolithic Qafzeh Cave, Israel: Indications for modern behavior. *J. Hum. Evol.*, *56*: 307–314.

Berland, S., Marie, A., Duplat, D. et al. (2011). Coupling proteomics and transcriptomics for the identification of novel and variant forms of mollusk shell proteins: A study with *P. margaritifera*. *Chembiochem*, *12*: 950–961.

Borrello, M. (2001). Vous avez dit "corail"? Annu. Soc. Suisse Préhistoire et d'Archéologie, 84: 191–196.

Bouzouggar, A., Barton, N., Vanhaeren, M. et al. (2007). 82,000-year-old shell beads from North Africa and implications for the origins of modern human behavior. *Proc. Natl. Acad. Sci. U.S.A.*, *104*: 9964–9969.

Brooks, A.S., Hare, P.E., Kokis, J.E. et al. (1990). Dating Pleistocene archeological sites by protein diagenesis in ostrich eggshell. *Science*, *248*: 60–64.

Brown, S., Higham, T., Slon, V. et al. (2016). Identification of a new hominin bone from Denisova Cave, Siberia using collagen fingerprinting and mitochondrial DNA analysis. *Sci. Rep.*, *6*: 23559.

Buckley, M., Walker, A., Ho, S.Y.W. et al. (2008). Comment on "Protein sequences from mastodon and *Tyrannosaurus rex* revealed by mass spectrometry." *Science*, *319*: 33.

Buckley, M., Warwood, S., van Dongen, B. et al. (2017). A fossil protein chimera; difficulties in discriminating dinosaur peptide sequences from modern cross-contamination. *Proc. R. Soc. B*, *284*: 20170544.

Cappellini, E., Jensen, L.J., Szklarczyk, D. et al. (2012). Proteomic analysis of a Pleistocene mammoth femur reveals more than one hundred ancient bone proteins. *J. Proteome Res.*, *11*: 917–926.

Cappellini, E., Gentry, A., Palkopoulou, E. et al. (2014). Resolution of the type material of the Asian elephant, *Elephas maximus* Linnaeus, 1758 (Proboscidea, Elephantidae). *Zool. J. Linn. Soc.*, *170*: 222–232.

Cappellini, E., Welker, F., Pandolfi, L., and Madrigal, J.R. (2019). Early Pleistocene enamel proteome sequences from Dmanisi resolve Stephanorhinus phylogeny. *Nature*, *574*: 103–107.

Carlson, S.J. (2016). The evolution of Brachiopoda. *Annu. Rev. Earth Planet. Sci.*, *44*: 409–438.

Carter, J.G. (ed.) (1989). *Skeletal Biomineralization: Patterns, Processes and Evolutionary Trends*, vol. 5. American Geophysical Union. https:/doi.org/10.1029/SC005.

Castiblanco, G.A., Rutishauser, D., Ilag, L.L. et al. (2015). Identification of proteins from human permanent erupted enamel. *Eur. J. Oral Sci.*, *123*: 390–395.

Chappell, J. and Shackleton, N.J. (1986). Oxygen isotopes and sea level. *Nature*, *324*: 137–140.

Chen, F., Welker, F., Shen, C.-C. et al. (2019). A late Middle Pleistocene Denisovan mandible from the Tibetan Plateau. *Nature*, *569*: 409–412.

Cleland, T.P., Schroeter, E.R., Zamdborg, L. et al. (2015). Mass spectrometry and antibody-based characterization of blood vessels from *Brachylophosaurus canadensis*. *J. Proteome Res.*, *14*: 5252–5262.

Cölfen, H. and Antonietti, M. (2005). Mesocrystals: Inorganic superstructures made by highly parallel crystallization and controlled alignment. *Angew. Chem. Int. Ed. Engl.*, *44*: 5576–5591.

Cordeiro, C.M.M. and Hincke, M.T. (2016). Quantitative proteomics analysis of eggshell membrane proteins during chick embryonic development. *J. Proteomics* 130: 11–25.

Coutu, A.N., Whitelaw, G., le Roux, P., and Sealy, J. (2016). Earliest evidence for the ivory trade in Southern Africa: Isotopic and ZooMS analysis of seventh–tenth century ad ivory from KwaZulu-Natal. *Afr. Archaeol. Rev.*, *33*: 411–435.

Demarchi, B., Hall, S., Roncal-Herrero, T. et al. (2016). Protein sequences bound to mineral surfaces persist into deep time. *eLife*, *5*: e17092.

d'Errico, F., Henshilwood, C., Vanhaeren, M., and van Niekerk, K. (2005). *Nassarius kraussianus* shell beads from Blombos Cave: Evidence for symbolic behaviour in the Middle Stone Age. *J. Hum. Evol., 48*: 3–24.

Dickinson, M.R., Lister, A.M., and Penkman, K.E.H. (2019). A new method for enamel amino acid racemization dating: A closed system approach. *Quat. Geochron., 50*: 29–46.

Drake, J.L., Mass, T., Haramaty, L. et al. (2013). Proteomic analysis of skeletal organic matrix from the stony coral *Stylophora pistillata*. *Proc. Natl. Acad. Sci. U.S.A., 110*: 3788–3793.

Emiliani, C. (1954). Depth habitats of some species of pelagic Foraminifera as indicated by oxygen isotope ratios. *Am. J. Sci., 252*: 149–158.

Emiliani, C. (1955). Mineralogical and chemical composition of the tests of certain Pelagic foraminifera. *Micropaleontology, 1*: 377–380.

Endo, K., Walton, D., Reyment, R.A., and Curry, G.B. (1995). Fossil intra-crystalline biomolecules of brachiopod shells: Diagenesis and preserved geo-biological information. *Org. Geochem., 23*: 661–673.

Faerman, M., Filon, D., Kahila, G. et al. (1995). Sex identification of archaeological human remains based on amplification of the X and Y amelogenin alleles. *Gene, 167*: 327–332.

Falini, G., Fermani, S., and Goffredo, S. (2015). Coral biomineralization: A focus on intra-skeletal organic matrix and calcification. *Semin. Cell Dev. Biol., 46*: 17–26.

Freeman, C.L., Harding, J.H., Quigley, D., and Rodger, P.M. (2011). Simulations of ovocleidin-17 binding to calcite surfaces and its implications for eggshell formation. *J. Phys. Chem. C, 115*: 8175–8183.

Fürst, S., Müller, K., Gianni, L. et al. (2016). Raman investigations to identify *Corallium rubrum* in Iron Age jewelry and ornaments. *Minerals, 6*: 56.

Gao, L., Tao, D., Shan, Y. et al. (2010). HPLC-MS/MS shotgun proteomic research of deer antlers with multiparallel protein extraction methods. *J. Chromatogr. B Analyt. Technol. Biomed. Life Sci., 878*: 3370–3374.

Gautron, J., Réhault-Godbert, S., Pascal, G. et al. (2011). Ovocalyxin-36 and other LBP/BPI/PLUNC-like proteins as molecular actors of the mechanisms of the avian egg natural defences. *Biochem. Soc. Trans., 39*: 971–976.

Goodfriend, G.A., Hare, P.E., and Druffel, E.R.M. (1992). Aspartic acid racemization and protein diagenesis in corals over the last 350 years. *Geochim. Cosmochim. Acta, 56*: 3847–3850.

Griffin, R.C., Chamberlain, A.T., Hotz, G. et al. (2009). Age estimation of archaeological remains using amino acid racemization in dental enamel: A comparison of morphological, biochemical, and known ages-at-death. *Am. J. Phys. Anthropol., 140*: 244–252.

Gupta, N.S. and Summons, R.E. (2011). Fate of chitinous organisms in the geosphere. In: *Chitin*, 133–151. Dordrecht: Springer.

Gutiérrez-Zugasti, F.I.G. (2011). The use of echinoids and crustaceans as food during the Pleistocene-Holocene Transition in Northern Spain: Methodological contribution and dietary assessment. *J. Island Coast. Archaeol., 6*: 115–133.

Haugen, J.-E., Sejrup, H.P., and Vogt, N.B. (1989). Chemotaxonomy of Quaternary benthic foraminifera using amino acids. *J. Foraminiferal Res., 19*: 38–51.

Hendy, E.J., Tomiak, P.J., Collins, M.J. et al. (2012). Assessing amino acid racemization variability in coral intra-crystalline protein for geochronological applications. *Geochim. Cosmochim. Acta, 86*: 338–353.

Hincke, M.T., Gautron, J., Tsang, C.P. et al. (1999). Molecular cloning and ultrastructural localization of the core protein of an eggshell matrix proteoglycan, ovocleidin-116. *J. Biol. Chem.*, *274*: 32915–32923.

Hodge, A.J., Petruska, J.A., and Bailey, A.J. (1965). The subunit structure of the tropocollagen macromolecule and its relation to various ordered aggregation states. In: *Structure and Function of Connective and Skeletal Tissue*, 31–41. London: Butterworths.

Immel, F., Gaspard, D., Marie, A. et al. (2015). Shell proteome of rhynchonelliform brachiopods. *J. Struct. Biol.*, *190*: 360–366.

Isowa, Y., Sarashina, I., Oshima, K. et al. (2015). Proteome analysis of shell matrix proteins in the brachiopod *Laqueus rubellus*. *Proteome Sci.*, *13*: 21.

Jackson, D.J., McDougall, C., Woodcroft, B. et al. (2010). Parallel evolution of nacre building gene sets in molluscs. *Mol. Biol. Evol.*, *27*: 591–608.

Jackson, D.J., Mann, K., Häussermann, V. et al. (2015). The *Magellania venosa* biomineralizing proteome: A window into brachiopod shell evolution. *Genome Biol. Evol.*, *7*: 1349–1362.

Jágr, M., Eckhardt, A., Pataridis, S., and Mikšík, I. (2012). Comprehensive proteomic analysis of human dentin. *Eur. J. Oral Sci.*, *120*: 259–268.

Jeong, C., Wilkin, S., Amgalantugs, T. et al. (2018). Bronze Age population dynamics and the rise of dairy pastoralism on the eastern Eurasian steppe. *Proc. Natl. Acad. Sci. U.S.A.*, *115*: E11248–E11255.

Jerardino, A. and Marean, C.W. (2010). Shellfish gathering, marine paleoecology and modern human behavior: Perspectives from cave PP13B, Pinnacle Point, South Africa. *J. Hum. Evol.*, *59*: 412–424.

Johnson, B.J., Miller, G.H., Fogel, M.L., and Beaumont, P.B. (1997). The determination of late Quaternary paleoenvironments at Equus Cave, South Africa, using stable isotopes and amino acid racemization in ostrich eggshell. *Palaeogeogr. Palaeoclimatol. Palaeoecol.*, *136*: 121–137.

Johnson, B.J., Fogel, M.L., and Miller, G.H. (1998). Stable isotopes in modern ostrich eggshell: A calibration for paleoenvironmental applications in semi-arid regions of southern Africa. *Geochim. Cosmochim. Acta*, *62*: 2451–2461.

Jonchère, V., Brionne, A., Gautron, J., and Nys, Y. (2012). Identification of uterine ion transporters for mineralisation precursors of the avian eggshell. *BMC Physiol.*, *12*: 10.

Jones, R.W. (2014). *Foraminifera and their Applications*. Cambridge University Press.

Jope, M. (1967). The protein of brachiopod shell – I. Amino acid composition and implied protein taxonomy. *Comp. Biochem. Physiol.*, *20*: 593–600.

Jope, M. (1977). Brachiopod shell proteins: Their functions and taxonomic significance. *Integr. Comp. Biol.*, *17*: 133–140.

Joubert, C., Piquemal, D., Marie, B. et al. (2010). Transcriptome and proteome analysis of *Pinctada margaritifera* calcifying mantle and shell: Focus on biomineralization. *BMC Genomics*, *11*: 613.

Kaufman, D.S. (2006). Temperature sensitivity of aspartic and glutamic acid racemization in the foraminifera Pulleniatina. *Quat. Geochronol.*, *1*: 188–207.

Kaye, T.G., Gaugler, G. and Sawlowicz, Z. (2008). Dinosaurian soft tissues interpreted as bacterial biofilms. *PLoS One*, *3*: e2808.

Kocot, K.M., Aguilera, F., McDougall, C. et al. (2016). Sea shell diversity and rapidly evolving secretomes: Insights into the evolution of biomineralization. *Front. Zool.*, *13*: 23.

Kucera, M. (2007). Planktonic foraminifera as tracers of past oceanic environments. In: *Developments in Marine Geology*, vol. *1* (eds. C. Hillaire-Marcel and A. De Vernal), 213–262. Elsevier.

Kunkel, J.G. and Jercinovic, M.J. (2013). Carbonate apatite formulation in cuticle structure adds resistance to microbial attack for American lobster. *Mar. Biol. Res.*, *9*: 27–34.

Lakshminarayanan, R., Valiyaveettil, S., Rao, V.S., and Kini, R.M. (2003). Purification, characterization, and in vitro mineralization studies of a novel goose eggshell matrix protein, ansocalcin. *J. Biol. Chem.*, *278*: 2928–2936.

Lowe, J.J. and Walker, M.J.C. (2014). *Reconstructing Quaternary Environments*. Routledge.

Mann, K. (2004). Identification of the major proteins of the organic matrix of emu (*Dromaius novaehollandiae*) and rhea (*Rhea americana*) eggshell calcified layer. *Br. Poult. Sci.*, *45*: 483–490.

Mann, K. (2015). The calcified eggshell matrix proteome of a songbird, the zebra finch (*Taeniopygia guttata*). *Proteome Sci.*, *13*: 29.

Mann, K. and Edsinger, E. (2014). The *Lottia gigantea* shell matrix proteome: Re-analysis including MaxQuant iBAQ quantitation and phosphoproteome analysis. *Proteome Sci.*, *12*: 28.

Mann, K. and Mann, M. (2013). The proteome of the calcified layer organic matrix of turkey (*Meleagris gallopavo*) eggshell. *Proteome Sci.*, *11*: 40.

Mann, K. and Siedler, F. (1999). The amino acid sequence of ovocleidin 17, a major protein of the avian eggshell calcified layer. *Biochem. Mol. Biol. Int.*, *47*: 997–1007.

Mann, K. and Siedler, F. (2004). Ostrich (*Struthio camelus*) eggshell matrix contains two different C-type lectin-like proteins. Isolation, amino acid sequence, and posttranslational modifications. *Biochim. Biophys. Acta*, *1696*: 41–50.

Manning, P.L., Morris, P.M., McMahon, A. et al. (2009). Mineralized soft-tissue structure and chemistry in a mummified hadrosaur from the Hell Creek Formation, North Dakota (USA). *Proc. Biol. Sci.*, *276*: 3429–3437.

Marean, C.W., Bar-Matthews, M., Bernatchez, J. et al. (2007). Early human use of marine resources and pigment in South Africa during the Middle Pleistocene. *Nature*, *449*: 905–908.

Marie, B., Marie, A., Jackson, D.J. et al. (2010). Proteomic analysis of the organic matrix of the abalone *Haliotis asinina* calcified shell. *Proteome Sci.*, *8*: 54.

Marie, B., Arivalagan, J., Mathéron, L. et al. (2017). Deep conservation of bivalve nacre proteins highlighted by shell matrix proteomics of the *Unionoida Elliptio complanata* and *Villosa lienosa*. *J. R. Soc. Interface*, **14**: 20160846.

Marin, F., Corstjens, P., de Gaulejac, B. et al. (2000). Mucins and molluscan calcification: Molecular characterization of mucoperlin, a novel mucin-like protein from the nacreous shell layer of the fan mussel *Pinna nobilis* (Bivalvia, Pteriomorphia). *J. Biol. Chem.*, *275*: 20667–20675.

Marin, F., Le Roy, N., and Marie, B. (2012). The formation and mineralization of mollusk shell. *Front. Biosci.*, *4*: 1099–1125.

Masters, P.M. (1986). Age at death determinations for autopsied remains based on aspartic acid racemization in tooth dentin: Importance of postmortem conditions. *Forensic Sci. Int.*, *32*: 179–184.

McDougall, C. and Degnan, B.M. (2018). The evolution of mollusc shells. *Wiley Interdiscip. Rev. Dev. Biol.*, *7*: e313.

Miller, G.H., Beaumont, P.B., Jull, A.J.T., and Johnson, B. (1992). Pleistocene geochronology and palaeothermometry from protein diagenesis in ostrich eggshells: Implications for the evolution of modern humans. *Philos. Trans. R. Soc. Lond. B Biol. Sci.*, *337* (1280): 149–157.

Miller, G., Magee, J., Smith, M. et al. (2016). Human predation contributed to the extinction of the Australian megafaunal bird *Genyornis newtoni* ~47 ka. *Nat. Commun.*, *7*: 10496.

Morel, J.-P., Rondi-Costanzo, C. and Ugolini, D. (2000). Corallo di ieri, corallo di oggi. Atti del Convegno di Ravello, Villa Rufolo, 13–15 dicembre 1996. Edipuglia.

Müller, P.J. (1984). Isoleucine epimerization in Quaternary planktonic foraminifera; effects of diagenetic hydrolysis and leaching, and Atlantic–Pacific intercore correlations. *Meteor. Forschungsergeb., Reihe C*, *38*: 25–47.

O'Connor, S., Solazzo, C., and Collins, M. (2015). Advances in identifying archaeological traces of horn and other keratinous hard tissues. *Stud. Conserv.*, *60*: 393–417.

Ogino, T., Ogino, H., and Nagy, B. (1985). Application of aspartic acid racemization to forensic odontology: Post mortem designation of age at death. *Forensic Sci. Int.*, *29*: 259–267.

Ohtani, S. (1994). Age estimation by aspartic acid racemization in dentin of deciduous teeth. *Forensic Sci. Int.*, *68*: 77–82.

Ohtani, S. and Yamamoto, T. (2010). Age estimation by amino acid racemization in human teeth. *J. Forensic Sci.*, *55*: 1630–1633.

Parker, G.J., Yip, J.M., Eerkens, J.W. et al. (2019). Sex estimation using sexually dimorphic amelogenin protein fragments in human enamel. *J. Archaeol. Sci.*, *101*: 169–180.

Porto, I.M., Laure, H.J., Tykot, R.H. et al. (2011). Recovery and identification of mature enamel proteins in ancient teeth. *Eur. J. Oral Sci.*, 119, Suppl *1*: 83–87.

Presslee, S., Wilson, J., Russell, D.G.D. et al. (2018). The identification of archaeological eggshell using peptide markers. *STAR: Sci. Technol. Archaeol. Res.*, *3*: 89–99.

Procopio, N., Chamberlain, A.T., and Buckley, M. (2017). Intra- and interskeletal proteome variations in fresh and buried bones. *J. Proteome Res.*, *16*: 2016–2029.

Ramos-Silva, P., Kaandorp, J., Huisman, L. et al. (2013). The skeletal proteome of the coral *Acropora millepora*: The evolution of calcification by co-option and domain shuffling. *Mol. Biol. Evol.*, *30*: 2099–2112.

Ramos-Silva, P., Kaandorp, J., Herbst, F. et al. (2014). The skeleton of the staghorn coral *Acropora millepora*: Molecular and structural characterization. *PLoS One*, *9*: e97454.

Reznikov, N., Bilton, M., Lari, L. et al. (2018). Fractal-like hierarchical organization of bone begins at the nanoscale. *Science*, **360**: eaao2189.

Rink, W.J. and Thompson, J.W. (2015). *Encyclopedia of Scientific Dating Methods*. Springer.

Ritz, S., Schütz, H.W., and Schwarzer, B. (1990). The extent of aspartic acid racemization in dentin: A possible method for a more accurate determination of age at death? *Z. Rechtsmed.*, *103*. 457–462.

Robbins, L.L. and Donachy, J.E. (1991). Mineral regulating proteins from fossil planktonic foraminifera. In: *Surface Reactive Peptides and Polymers*, vol. *444*, 139–148. American Chemical Society.

Robbins, L.L., Toler, S.K., and Donachy, J.E. (1993). Immunological and biochemical analysis of test matrix proteins in living and fossil foraminifers. *Lethaia*, *26*: 269–273.

Rodríguez-Navarro, A.B., Marie, P., Nys, Y. et al. (2015). Amorphous calcium carbonate controls avian eggshell mineralization: A new paradigm for understanding rapid eggshell calcification. *J. Struct. Biol.*, *190*: 291–303.

Romano, P., Fabritius, H., and Raabe, D. (2007). The exoskeleton of the lobster *Homarus americanus* as an example of a smart anisotropic biological material. *Acta Biomater.*, *3*: 301–309.

Sabbatini, A., Bédouet, L., Marie, A. et al. (2014). Biomineralization of *Schlumbergerella floresiana*, a significant carbonate-producing benthic foraminifer. *Geobiology*, *12*: 289–307.

Sadat-Shojai, M., Khorasani, M.-T., Dinpanah-Khoshdargi, E., and Jamshidi, A. (2013). Synthesis methods for nanosized hydroxyapatite with diverse structures. *Acta Biomater.*, *9*: 7591–7621.

Saitta, E.T., Liang, R., Lau, C.Y. et al. (2019). Cretaceous dinosaur bone contains recent organic material and provides an environment conducive to microbial communities. *eLife*, *8*: e46205.

Sakalauskaite, J., Andersen, S., Biagi, P. et al. (2019). "Palaeoshellomics" reveals the use of freshwater mother-of-pearl in prehistory. *eLife*, *8*: e45644.

San Antonio, J.D., Schweitzer, M.H., Jensen, S.T. et al. (2011). Dinosaur peptides suggest mechanisms of protein survival. *PLoS One*, *6*: e20381.

Sawafuji, R., Cappellini, E., Nagaoka, T. et al. (2017). Proteomic profiling of archaeological human bone. *R. Soc. Open Sci.*, *4*: 161004.

Schroeter, E.R., DeHart, C.J., Cleland, T.P. et al. (2017). Expansion for the *Brachylophosaurus canadensis* collagen I sequence and additional evidence of the preservation of Cretaceous protein. *J. Proteome Res.*, *16*: 920–932.

Skeates, R. (1993). Mediterranean coral: Its use and exchange in and around the Alpine region during the later Neolithic and Copper Age. *Oxf. J. Archaeol.*, *12*: 281–292.

Solazzo, C., Wadsley, M., Dyer, J.M. et al. (2013). Characterisation of novel α-keratin peptide markers for species identification in keratinous tissues using mass spectrometry. *Rapid Commun. Mass Spectrom.*, *27*: 2685–2698.

Solazzo, C., Fitzhugh, W., Kaplan, S. et al. (2017). Molecular markers in keratins from Mysticeti whales for species identification of baleen in museum and archaeological collections. *PLoS One*, *12*: e0183053.

Stapane, L., Le Roy, N., Hincke, M.T., and Gautron, J. (2019). The glycoproteins EDIL3 and MFGE8 regulate vesicle-mediated eggshell calcification in a new model for avian biomineralisation. *J. Biol. Chem.*, *294*: 14526–14545.

Stathoplos, L. and Tuross, N. (1991). Mineral-associated proteins from modern planktonic foraminifera. In: *Mechanisms and Phylogeny of Mineralization in Biological Systems*, 17–21. Tokyo: Springer.

Stathoplos, L. and Tuross, N. (1994). Proteins and DNA from modern planktonic foraminifera. *J. Foraminiferal Res.*, *24*: 49–59.

Stewart, N.A., Molina, G.F., Issa, J.P.M. et al. (2016). The identification of peptides by nanoLC-MS/MS from human surface tooth enamel following a simple acid etch extraction. *RSC Adv.*, *6*: 61673–61679.

Stone, A.C., Milner, G.R., Pääbo, S., and Stoneking, M. (1996). Sex determination of ancient human skeletons using DNA. *Am. J. Phys. Anthropol.*, *99*: 231–238.

Stringer, C.B., Finlayson, J.C., Barton, R.N.E. et al. (2008). Neanderthal exploitation of marine mammals in Gibraltar. *Proc. Natl. Acad. Sci. U.S.A.*, *105*: 14319–14324.

Taviani, M. (1997). L'uomo e il corallo. *Ori delle Alpi*: 150–152.

Tomiak, P.J., Penkman, K.E.H., Hendy, E.J. et al. (2013). Testing the limitations of artificial protein degradation kinetics using known-age massive Porites coral skeletons. *Quat. Geochronol.*, *16*: 87–109.

von Holstein, I.C.C., Ashby, S.P., van Doorn, N.L. et al. (2014). Searching for Scandinavians in pre-Viking Scotland: Molecular fingerprinting of Early Medieval combs. *J. Archaeol. Sci.*, *41*: 1–6.

Wadsworth, C. and Buckley, M. (2014). Proteome degradation in fossils: Investigating the longevity of protein survival in ancient bone. *Rapid Commun. Mass Spectrom.*, *28*: 605–615.

Wadsworth, C. and Buckley, M. (2018). Characterization of proteomes extracted through collagen-based stable isotope and radiocarbon dating methods. *J. Proteome Res.*, *17*: 429–439.

Wadsworth, C., Procopio, N., Anderung, C. et al. (2017). Comparing ancient DNA survival and proteome content in 69 archaeological cattle tooth and bone samples from multiple European sites. *J. Proteomics*, *158*: 1–8.

Warinner, C., Rodrigues, J.F.M., Vyas, R. et al. (2014a). Pathogens and host immunity in the ancient human oral cavity. *Nat. Genet.*, *46*: 336–344.

Warinner, C., Hendy, J., Speller, C. et al. (2014b). Direct evidence of milk consumption from ancient human dental calculus. *Sci. Rep.*, *4*: 7104.

Wehmiller, J.F. (1980). Intergeneric differences in apparent racemization kinetics in mollusks and foraminifera: Implications for models of diagenetic racemization. In: *Biogeochemistry of Amino Acids* (eds. P.E. Hare, T.C. Hoering and K. King Jr.), 341–345. New York: Wiley.

Wehmiller, J. and Hare, P.E. (1971). Racemization of amino acids in marine sediments. *Science*, *173*: 907–911.

Weiner, S. and Traub, W. (1986). Organization of hydroxyapatite crystals within collagen fibrils. *FEBS Lett.*, *206*: 262–266.

Welker, F., Collins, M.J., Thomas, J.A. et al. (2015a). Ancient proteins resolve the evolutionary history of Darwin/'s South American ungulates. *Nature*, *522*: 81–84.

Welker, F., Soressi, M., Rendu, W. et al. (2015b). Using ZooMS to identify fragmentary bone from the late Middle/Early Upper Palaeolithic sequence of Les Cottes, France. *J. Archaeol. Sci.*, *54*: 279–286.

Welker, F., Hajdinjak, M., Talamo, S. et al. (2016). Palaeoproteomic evidence identifies archaic hominins associated with the Châtelperronian at the Grotte du Renne. *Proc. Natl. Acad. Sci. U.S.A.*, *113*: 11162–11167.

Welker, F., Smith, G.M., Hutson, J.M. et al. (2017). Middle Pleistocene protein sequences from the rhinoceros genus *Stephanorhinus* and the phylogeny of extant and extinct Middle/Late Pleistocene Rhinocerotidae. *PeerJ*, *5*: e3033.

Wu, T.M., Rodriguez, J.P., Fink, D.J. et al. (1995). Crystallization studies on avian eggshell membranes: Implications for the molecular factors controlling eggshell formation. *Matrix Biol.*, *14*: 507–513.

4
Chiral Amino Acids: Geochronology and Other Applications

4.1 Dating the Quaternary (Pleistocene and Holocene)

4.1.1 Palaeoclimate and Coastal Studies

The Quaternary – from 2.6 million years ago to the present day – is the age in which glacial–interglacial cycles become increasingly intense and the temperature shifts more acute. The measurement of the ratio of the stable oxygen isotopes ^{18}O to ^{16}O in marine microfossils contained within deep-ocean sediments (Emiliani, 1954; Chappell and Shackleton, 1986) shows that lighter $\delta^{18}O$ values represent warmer temperatures and heavier $\delta^{18}O$ values reflect colder intervals. Each 'isotopic' stage is given a number (odd numbers for the interglacials and even numbers for the glacial periods), based on a 'count from the top' principle, with MIS (Marine oxygen Isotope Stage) 1 representing the present interglacial, or Holocene. The duration of each stage is variable throughout the Quaternary. While 41 ka cycles dominated the earlier part of the Pleistocene, about 650 ka ago, increased-amplitude climatic oscillations began (Figure 4.1).

These multiple cold/warm cycles deeply altered the landscape, each cycle being characterized by advances and retreats of the ice sheets, expansion and contraction of periglacial areas, and variations in weathering rates, pedogenic processes, river regimes and sea level. These phenomena produced a rich and complex record of landforms, sediments and biological remains, which can be used to reconstruct palaeoenvironmental conditions. (See Lowe and Walker (2014) for an excellent overview.)

One of the key questions, since the 1960s at least, relates to whether a direct link can be made between the terrestrial record and the global MIS

Amino acids and Proteins in Fossil Biominerals: An Introduction for Archaeologists and Palaeontologists, First Edition. Beatrice Demarchi.
© 2020 John Wiley & Sons Ltd. Published 2020 by John Wiley & Sons Ltd.

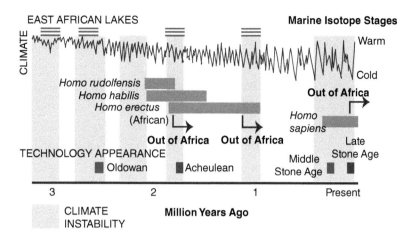

Figure 4.1 Coeval events of climatic instability and human evolution during the Quaternary in Africa. Figure adapted and simplified from Campisano (2012) and Slezak (2015).

signal. Ideally, this would be achieved by the dating of specific geological features, indicators of cold/warm stages, which would thus enable their association to an MIS and, consequently, allow us to build a global chronostratigraphy for Quaternary climatic changes and to better understand the effect of these changes on the environment. Some of the best places to target for this purpose are coastal environments, due to their extreme sensitivity to climatic variations which cause changes in sea level.

Quaternary global changes in ocean volume were predominantly caused by glacial cycles (glacio-eustasy). The building of ice sheets extracts water from the oceans, while subsequent melting of the ice during warm intervals releases vast volumes of water into the ocean (Murray-Wallace and Woodroffe, 2014). Warm stages are thus usually associated with higher sea levels. However, on larger time scales (tens to hundreds of millions of years), other factors contribute to eustatic changes, not only those such as sea-floor spreading and the filling and drying of large basins (e.g. the Mediterranean Sea), but also tectonic events with global significance, such as the closure of the Isthmus of Panama. Isostatic land uplift and subsidence are caused by a variety of tectonic processes, operating at different temporal and spatial scales and affecting the local radius of the solid Earth (Peltier, 1998). The major cause of sea level variation in areas of former glaciation is glacio-isostasy: during glaciations, the weight of ice on continental shelves and the rebound of the crust, which follows the melting of the ice, can produce substantial changes in relative sea level. This glacio-isostatic signal changes as a function of distance from the ice sheet – while it dominates near the centres of glaciation, closer to the margins it may become comparable to the rise resulting from the eustatic increase in ocean volume. A further complication is caused by the sea floor adjusting as the weight from the meltwater is added (hydro-isostasy), which causes a redistribution of water and a consequent variation in sea level. The combination of isostasy and eustasy can produce either a rise or fall in relative sea level.

The typical consequence of sea level change is the availability of new dry land during glacial periods, which then becomes re-submerged during deglaciation. It has been demonstrated that after the Last Glacial Maximum (LGM, around 26–19 ka BP), the global sea level was 125 ± 5 m below the present datum, while during the Last Interglacial (around 130–116 ka BP) it was 6 m higher than present (Chappell and Shackleton, 1986; Fleming et al., 1998; Lambeck and Chappell, 2001). It has been estimated that the percentage of dry land exposed was 10% higher during the last glaciation than it is today (Bailey, 2004).

Raised beaches are probably the most significant and visible coastal geological feature. These formed when sea level highstands reached or exceeded the present sea level. As such, many fossil shorelines can today appear to have been 'raised' (this name can be misleading, because fossil shorelines higher than present sea level could have been produced by the combination of eustatic and isostatic changes, both 'positive' and 'negative' (Haslett 2009)). The height of the raised beaches above present sea level can potentially be used to calibrate past sea transgressions. Raised beaches can be correlated with the interglacial stages within the oxygen isotope record by means of direct dating. Uranium series dating (Bender et al. 1979; Hillaire-Marcel et al. 1986; Li et al. 1989; Bard et al. 1990), luminescence (Huntley et al. 1985), electron spin resonance (Rutter et al. 1990; Pirazzoli et al. 2004) and radiocarbon (Zwartz et al. 1998) have been most commonly applied. Please see Rink and Thompson (2015) for a comprehensive review of the different techniques for dating raised beaches. Amino acid racemization has been frequently and successfully applied to the development of geochronological frameworks for Quaternary beaches in many areas of the world (see Figure 4.7 and a review in Wehmiller, 2015a).

4.1.2 *Human Evolution Against a Backdrop of Quaternary Climate Change*

Despite considerable efforts, our picture of human evolution during the Quaternary can be defined as sketchy at best. Questions regarding the dispersal of human populations, the development of technology and the degree of adaptation to highly variable environments, are still open to debate. There is a strong case to be made for correlating major transitions in human evolution with palaeoenvironmental shifts (see for example Figure 4.1), but the fossil evidence is so scattered and diverse and the depositional processes so poorly understood that building a general interpretative framework has proven difficult. A major problem identified by many authors (see Mcbrearty and Brooks, 2000, for example) is the lack of convincing chronological information for many archaeological and geological sites and sequences. If this were available, it would dramatically improve our understanding of the past, forming a solid basis for the correlation of contemporary events across space and for the tracing of evolutionary patterns across time.

The numerical dates published over the last 60 years for key archaeological sites are often associated with large uncertainties and may need re-evaluation (see a discussion in Millard, 2008). Part of the reason is methodological: some dating methods are more reliable than others, because they have been tried and tested over decades, i.e. more studies have been carried out and have identified different problems potentially affecting the technique. This is the case, for example, with radiocarbon dating. Newer techniques can suffer from a number of unidentified problems, which are addressed only as the development of the technical procedures progresses. Furthermore, methodological issues imply that the direct comparison between dates obtained by two different techniques, or by using 'old' and 'new' analytical protocols of the same technique, or on two different substrates (e.g. radiocarbon dates obtained on coeval bone and charcoal), is not to be recommended.

The use of harsh pretreatments for charcoal (the ABOx method) and of ultrafiltration for bone collagen has significantly improved the radiocarbon chronology of many sites in the past 15 years (Higham et al., 2006, 2014; Wood et al., 2013), and evidence suggests that equally harsh pretreatments (i.e. a strong acid–base–acid wash) are needed to improve the dating of samples of vegetable matter (e.g. seeds) and sediments, which are prone to contamination from humic acids and other compounds (Briant et al., 2018). Sediments are often the only substrates available for dating terrestrial sequences that need to be correlated to the Marine Isotope Stage record and fine tuned to provide chronologies between 30–50 ka BP, i.e. at the limit of the resolution for radiocarbon dating. Incidentally, this is the time of the last glacial, in which important extinctions, of both megafauna and human lineages (Neanderthals, Denisovans) occurred.

Another improvement in the scientific dating of the Middle and Upper Pleistocene has been the incorporation of stratigraphic, chronometric and other *a priori* information into Bayesian models (Ramsey, 2009). The use of Bayesian statistics has brought on a veritable revolution of radiocarbon dating, and has allowed the dating of sites which lie at the very boundary of its temporal span, e.g. the Middle to Upper Palaeolithic transition in Eurasia. However, the use of Bayesian methods is not a panacea, and controversies still exist on the dating of some of these key sites: for example, ^{14}C measurements taken from mollusc shells from the same site (Ksâr 'Akil, Lebanon), calibrated and modelled using Bayesian statistics in two different studies, provided different ages (Douka et al., 2013, 2015; Bosch et al., 2015a, 2015b).

Overall, the evolution of radiocarbon dating as a technique provides an important cautionary tale for all chronological methods: if a technique is perceived to have 'issues', then these issues can be seen as confounding factors by the end users and, consequently, simply ignored. As a result, the practice of choosing the dates which best suit the preferred evolutionary scheme or archaeological interpretation was (is) sadly widespread. Re-evaluation of old dates in light of new statistical models and analytical protocols, and a more critical approach to the dating process, will certainly provide a more solid background for the use of numerical dates to inform archaeological inferences.

However, two other problems still hamper the dating process for the Quaternary period: limited time depth and substrate specificity of most geochronological techniques. Few techniques, if any, are able to span the whole Quaternary (Figure 4.2), most of them being limited to very specific time ranges and suffering from a decrease in resolution/accuracy when pushed to their limits. This is particularly the case for the dating of the Lower and Middle Palaeolithic, i.e. the period between 3 Ma and 40 ka BP, in which the major transitions in human evolution and dispersals occurred. Another limitation is that scientific dating techniques can be used to date only *some* types of material (e.g. quartz grains for luminescence). A review of scientific geochronological tools can be found in Walker (2005) and in the recent *Encyclopaedia of Scientific Dating Methods* (Rink and Thompson, 2015). Here we will concern ourselves with the role of protein diagenesis dating (or amino acid racemization dating) in improving the geochronology of the Quaternary.

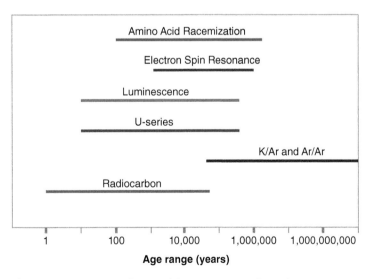

Figure 4.2 Time span of some of the most commonly used Quaternary geochronological techniques. Note the logarithmic scale.

4.2   Principles of AAR Dating

Amino acid racemization (AAR) as a dating technique has the potential to improve the geochronological control for the last 3 Ma. It has a distinctively large range of applications, spanning the whole Quaternary in subarctic and temperate areas and being suitable for the dating of fossils that are widespread in many areas of the world, such as mollusc shells. The main idea behind AAR dating is simple: following the death of an organism (or, more accurately, after cessation of protein turnover), the original organic component undergoes degradation (as described in Chapter 2) and the extent of protein breakdown increases with the age of the fossil (Hare and Abelson, 1968). If the extent of degradation can be quantified, this value can be linked to the time elapsed since death, provided that all factors affecting the rate of racemization are known or can be modelled/inferred. Biominerals are especially suited to AAR dating since they generally display good preservation (thanks to their exceptional mechanical properties) and are ubiquitous in both the archaeological and geological fossil records.

The basic principles of AAR geochronology are shown in Figure 4.3 and explained here.

1   Racemization is a post mortem spontaneous reaction, involving reversible interconversion between two enantiomers of a single amino acid, the D and L forms.

2   L-amino acids are present in living organisms, while D-amino acids are formed post mortem by racemization.

3   The extent of racemization can be measured as the ratio between the D and L forms detected (by chromatography) in a fossil sample: this is the D/L value.

4   The D/L value for the amino acid isoleucine (one of the most utilized for the building of chronological frameworks) is known as the A/I value, since L-isoleucine converts into D-alloisoleucine; in this case, Ile racemization is more correctly referred to as epimerization (Chapter 2).

5   The D/L (or A/I) value yields an estimate of the time elapsed since the death of the organism: older fossils will have higher D/L values (closer to 1), while living organisms should have a D/L of 0, because they contain L enantiomers only.

6   It is possible to analyse two fractions of amino acids: the free (FAA) and total hydrolysable (THAA) fractions (although most studies usually analyse only the latter). Free amino acids are those naturally present in the sample, since they are released by hydrolysis of the protein chain; total amino acids are all amino acids present in the sample, both free and protein bound. Bound amino acids are released by an acid hydrolysis step, performed at high temperature (bearing in mind that this step may not be able to release all bound amino acids, particularly

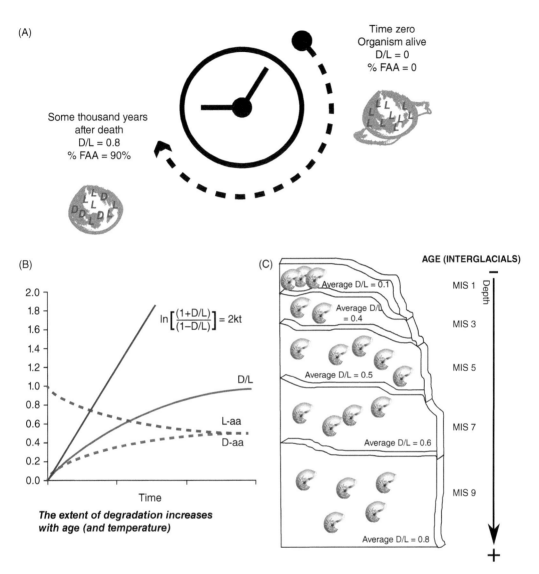

Figure 4.3 (A) All amino acids are in the L form in living organisms (here, a live mollusc with its shell). After death, the organisms decay. Soft tissues disappear rapidly, but proteins are preserved in the calcified shell; the racemization reaction begins and over time more D enantiomers are accumulated. The rate of racemization depends on a range of factors (primarily, temperature). In temperate environments, after several hundred thousand years the reaction will reach dynamic equilibrium, that is the forward (L → D) and backward (D → L) rate constants will be the same and there will be an approximately equal concentration of D and L enantiomers (D/L = 1). (B) If the reactions were happening in a closed system of free amino acids in solution, they would follow first-order reversible kinetics, and it would be possible to calculate the rate of reaction and time since death of the organism. (C) In reality, the reaction pathways are complex, and the most useful application for AAR is relative dating: in this theoretical example, shell-bearing sediments accumulated during odd-numbered MISs (i.e. interglacials) can be placed on a relative timescale by measuring the average D/L values on multiple shells (note that all shells belong to the same taxon).

if these are trapped in complex high molecular weight molecules). Therefore, for each sample we can obtain a FAA D/L and a THAA D/L value, provided that the proteins analysed display a closed-system behaviour (in which the closed system has been isolated by bleaching or diagenesis, see Chapter 1).

The D/L value can be used in two main ways:

1 to obtain relative age information, particularly useful for ordering sedimentary stratigraphic layers containing fossil biominerals, from youngest to oldest (aminostratigraphy);
2 to predict the age since death of a fossil, when the rate of racemization is known (aminochronology) – this can be obtained using kinetic models (using artificial degradation experiments at high temperature) in combination with calibration of D/L values (using additional numerical techniques; e.g. OSL, U-series, radiocarbon).

We will examine how to measure the D/L values first, and then move on to consider the principles and applications of aminostratigraphy and aminochronology.

4.3   Measuring D/L Values

4.3.1   FAA and THAA Fractions

As we have seen, in closed system biominerals there are two fractions of amino acids: THAA and FAA. FAAs are released by natural hydrolysis and are more mobile than peptide chains. As such, if we were to consider a 'real' subfossil biomineral, FAAs would represent the fraction that tends to leach out of the mineral and to be lost in the ground. However, in a closed system isolated by bleaching or diagenesis (see Chapter 1), FAAs are trapped within the crystals and can therefore provide an estimate of the extent of degradation of the protein. THAA D/L values reflect a combination of the racemization of the amino acid residues while peptide bound (usually as the highly racemized N-terminal amino acid, see Figure 2.5), followed by racemization as a free residue. As a result, FAA D/L values tend to be higher than THAA D/L values, as the latter incorporate the D/L values of all internally bound amino acids. In order to analyse FAA, the bleached biomineral powder is demineralized gently using a dilute acid (e.g. 2 M HCl, cold) in order to release the amino acids from the intracrystalline space. On the contrary, THAAs include both FAA and the residues that are still peptide bound. Therefore, demineralization and acid hydrolysis are carried out with a single step at high temperature (e.g. 7 M HCl, 24 hours at 110 °C).

4.3.2   *Chromatography*

In order to obtain a measurement of the D/L values (one for each amino acid), the mixture of FAA and THAA found in fossils needs to be separated into its components, that is the enantiomers of all amino acids. Standard chromatographic techniques are traditionally used for this purpose (Miller and Brigham-Grette, 1989; Wehmiller and Miller, 2000). Chromatography is based on the differential elution of an analyte over time, according to the chemical affinities of its components with a mobile/stationary phase. Reviewing the principles and applications of chromatography is beyond our scope, so the interested reader is directed towards a recent manual on chromatography (Vitha, 2016), while we briefly summarize the main types of chromatographic techniques used in AAR studies.

Ion-exchange liquid chromatography (IEC) was most commonly used in the early days of AAR dating (Hare and Mitterer, 1969; Hare et al., 1980). This method can be successfully applied to quantify the amino acids present in a mixture extracted from fossil samples, but it is not able to separate enantiomeric pairs. Therefore only one D/L value is generated, as only the two common diastereomers of isoleucine (D-alloisoleucine and L-isoleucine) can be resolved, due to their different properties. In order to detect the enantiomers of multiple amino acids, gas chromatography can be performed successfully (Kvenvolden et al., 1973). Due to the nonvolatility of amino acids, the technique usually involves a laborious sample pretreatment, which may have an impact on the number of samples that can be processed. However, the pretreatment also has the advantages both of desalting the sample and yielding baseline resolution of almost all amino acids, as well as of improving the life span of the column (by up to 10 years, running 50 samples per week with at least three injections per sample; J. Wehmiller, personal communication, February 2019).

In recent decades, the method of liquid chromatography that has been typically applied to amino acid separation is high-pressure liquid chromatography (HPLC), in which high pressure forces the analyte through a column. Reverse-phase HPLC (RP-HPLC) combines the high sensitivity and the minimal sample pretreatment procedures typical of HPLC with the ability to resolve multiple enantiomeric pairs. Since the two enantiomers of single amino acids exhibit identical chemical and physical properties, a derivatization step is necessary so that they can be separated by chromatography and detected by a fluorescence detector. Derivatization is usually performed with a chiral thiol compound (e.g. OPA-IBLC) which reacts with the amino acids to form highly fluorescent compounds (Fitznar et al., 1999). However, the D and L enantiomers differ in their fluorescence intensity and, as such, the D/L values must be calculated by calibrating the signal with standards of known concentration.

The RP-HPLC method currently in use in all AAR laboratories was developed by Kaufman and Manley (1998). A nonpolar stationary phase and a polar mobile phase are used to separate the enantiomers which are derivatized precolumn by mixing a solution volume of 2 µL of sample with 2.2 µL of derivatizing reagent. The reagent is 260 mM *N*-isobutyryl-L-cysteine (IBLC), 170 mM *o*-phthaldialdehyde (OPA) in 1 M potassium borate buffer, pH 10.4. The resulting enantiomeric derivatives are then separated on a C18 column at 25 °C, using a linear gradient of three solvents: sodium acetate buffer (23 mM sodium acetate trihydrate, 1.3 µM Na_2EDTA, 1.5 mM sodium azide, adjusted to pH 6.00 ± 0.01), methanol and acetonitrile, keeping the elution time below 120 minutes.

The elution time of each stereoisomer is a function of its mass, structure and hydrophobicity. The separated amino acids are then identified by the fluorescence detector and each elution recorded as separate peaks on a chromatogram. Under these conditions the area under each peak is directly proportional to the concentration of each amino acid, which is then normalized to the internal standard (L-h-Arg). The technique allows detection in the subpicomole range: Kaufman and Manley (1998) found that the limit of quantifiable detectability was as low as 0.1 pmol. The L and D isomers of the following amino acids are routinely analysed: L-Asp, D-Asp, L-Glu, D-Glu, L-Ser, D-Ser, L-Thr, L-His, Gly (which often co-elutes with D-Thr and D-His), L-Arg, D-Arg, L-Ala, L-h-Arg (internal standard), D-Ala, L-Tyr, D-Tyr, L-Val, L-Met, D-Met, D-Val, L-Phe, L-Ile, D-Phe, L-Leu, D-aIle, D-Leu (Figure 4.4). During preparative hydrolysis both asparagine and glutamine undergo rapid irreversible deamidation to aspartic acid and glutamic acid, respectively (Hill, 1965). It is therefore not possible to distinguish between the acidic amino acids and their derivatives, and they are reported together as Asx and Glx.

A number of advantages result from the use of this RP-HPLC method.

1 The amount of sample required is very small, around 2 mg (of bleached biomineral) or less. This can prove crucial for the analysis of fossil material, which is often scarce or too precious to destructively analyse in large amounts. In order to control for inter-sample variability, multiple samples should be analysed (the larger sample size required for GC, i.e. 300 mg, is problematic for small samples, but is conversely more representative of larger samples, e.g. thick shell layers; J. Wehmiller, personal communication, February 2019).

2 The high automation of the system allows for maximum efficiency of the analysis, increasing the number of samples which can be processed, therefore improving the statistical significance of a dataset and allowing the estimate of variability between multiple injections of the same sample, multiple preparations of the same sample, as well as multiple samples from the same sedimentary layer (see Miller and Brigham-Grette, 1989; Wehmiller and Miller, 2000; Simonson et al., 2013, for example).

3 It is possible to detect diagenetically compromised samples based on the lack of expected covariance between the D/L values of multiple

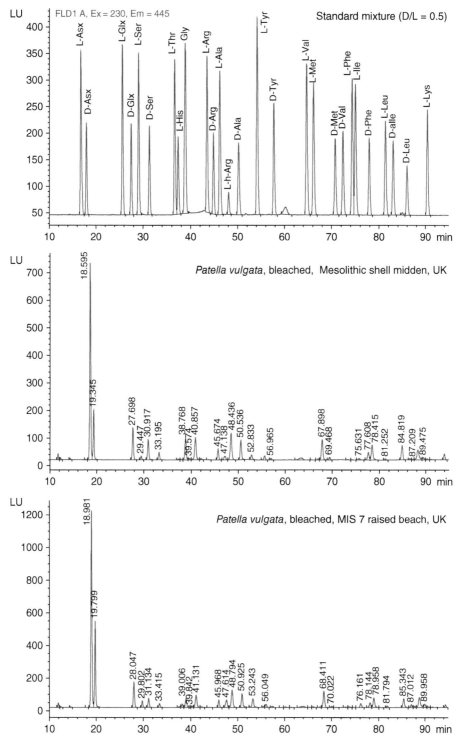

Figure 4.4 Typical RP-HPLC chromatograms (NEaar laboratory, York, UK). From top to bottom: standard mixture of chiral amino acids (D/L = 0.5); bleached powders of a *Patella vulgata* shell, from a Mesolithic shell midden in Scotland, i.e. about 9000 years old; bleached powders of a *Patella vulgata* shell from a raised beach in Northern England dating to the penultimate Interglacial, about 250 000 years ago (MIS 7).

amino acids and/or between the THAA and FAA values for the same amino acid, or on the overabundance of easily decomposed amino acids (typically, L-Ser, which indicates modern contamination, or D-Ser, which indicates bacterial peptidoglycan contamination).

A major disadvantage of the method is the poor resolution of the L-Ile/D-aIle peaks, while isoleucine is very well resolved by IE chromatography. A recent study by Whiteacre et al. (2017) used paired analyses of eggshell, mollusc shell and whole-rock samples (which has become an important substrate for AAR in the past decades, see Hearty and Kaufman, 2000, for example) by RP-HPLC and IE, with the purpose of converting IE A/I values to equivalent RP-HPLC ratios for Asx, Glx, Ala, Val and Ile. While the uncertainties associated with this conversion are ±24–41% depending on the amino acid, this is the first time results from the two methods can be integrated, increasing the size of AAR datasets and improving its potential. We note here that excellent specificity and a reduction in run times, compared to the RP-HPLC method, can be achieved by exploiting recent advances in chromatographic techniques, such as ultra-high pressure liquid chromatography (UHPLC). A method for UHPLC analysis of chiral amino acids is described in the PhD thesis of M.K. Crisp (2013).

4.4 Factors Affecting D/L Values

Generally speaking, whether the age information provided by the D/L values is accurate or not depends strongly on (a) the type of biomineral (i.e. its open-system or closed-system behaviour with regard to protein diagenesis, as we have seen in Chapters 1 and 2); (b) differences between types of biominerals or between different taxa; and (c) the burial environment (mixing of sediments, thermal effects). We will briefly discuss each of these, bearing in mind that while careful sampling and pretreatment of samples are able to reduce the uncertainty associated with D/L values, some of this variability cannot be eliminated.

4.4.1 Closed Versus Open Systems

If the proteins are retained within an open system, reliable D/L values cannot be obtained. The reason is simple: there is no way of controlling for modern contamination and, above all, for differential loss of amino acids from the system, which will skew the D/L values because the original degradation signature will not be maintained. Most biominerals behave as open systems, with the exception of ratite eggshell (and, even in this case, recent work has shown that the organic matrix is not fully intracrystalline: Crisp et al., 2013).

However, as highlighted previously, diagenesis effectively isolates a stable fraction of proteins, especially in older (pre-Holocene) samples. For younger samples, a bleaching pretreatment may be helpful for ensuring the analysis of proteins that do behave as a closed system. However, this is not possible for all substrates, the most important exception being bone. As seen in Chapters 1 and 3, bone does not retain a classically defined intracrystalline fraction. However, large crystal aggregates are present, which are resistant to NaOCl treatment and retain well preserved DNA (Salamon et al., 2005). This suggests that bleached bone may yet become a useful substrate for AAR geochronology in the future.

With regard to bone, AAR saw its fortunes sink after a decade-long debate on the Asx racemization dates obtained on the skeletons of so-called 'Paleoindians' from a range of sites in California. The first AAR dates had been obtained by Bada and co-workers in the mid-1970s (Bada and Protsch, 1973; Bada et al., 1974) using Asp racemization rates calibrated on the key specimen of the Laguna Skull (dated by conventional radiocarbon to 15 000–17 000 years BP). This calibrated Asp rate was assumed to conform to linear kinetics and used to establish a relationship between time and D/L values, and to extend the 'calibration curve' back in time so that remains beyond the range of radiocarbon could be dated. This method seemed to indicate that humans had been present in California 70 000 years ago! In fact, later studies found that these remains were mid-Holocene in age (Taylor et al., 1983, 1985), and indeed the skeletons *looked* mid-Holocene (Dincauze, 1984). How did such a blunder occur? In an unfortunate combination of events, the radiocarbon date used to calibrate the racemization rate turned out to be wrong – using the radiocarbon dates obtained at a later date by AMS (accelerator mass spectrometry) for calibration resulted in less wild AAR estimates (Bada, 1985). The issue was not only with the radiocarbon dates, but also with the assumptions underlying the mechanisms of Asx racemization. Although the amino acid sequence of collagen appears to be quite favourable to Asx racemization, since Asp and Asn occur N terminal to Gly (a position in which Asx is especially prone to undergo rapid racemization), the helical structure of the molecule prevents this racemization from occurring (van Duin and Collins, 1998; Collins et al., 1999). Measured Asx D/L values therefore reflect: (a) initially, the low amount of Asx that is not bound in the triple helices (either the terminal frayed end of the collagen, the telopeptides, or the NCPs); (b) the gelatinization of collagen, which allows collagen-bound Asx residues to racemize rapidly; and (c) the rapid loss of soluble collagen (Collins et al., 2009). This three-pool model of variable Asx racemization rates is certainly far removed from any linear kinetic models. It follows that AAR aminostratigraphy and aminochronology on bone are not reliable applications of the technique. However, it is rather unfortunate that this 'Paleoindians' episode has somewhat stained the reputation of a geochronological technique which has been shown to be useful and reliable in hundreds of studies using shell, eggshell, foraminifera, ostracods and brachiopods, among others.

4.4.2   Species Effect

Differences in the observed racemization rates between different biominerals, different genera (e.g. of mollusc shells) but also between species within the same genus have been reported since the 1980s (e.g. Lajoie et al., 1980). As a result, it is not possible to directly translate the diagenesis patterns observed in a molluscan species, for example, in order to interpret those observed in another. In fact, taxonomic uncertainty is one of the main contributors to D/L variability (Miller and Brigham-Grette, 1989). The recommendation was therefore to use monospecific samples for any dating applications, an approach which is still being used today, not least because subsequent work has shown that the 'species effect' persists even within the closed system (Penkman et al., 2008; Demarchi et al., 2014). This may seem obvious in light of the data on the complexity of biomineralized proteomes presented in Chapter 3, but, at the time of discovery of the species effect, the mechanisms of protein-driven mineralization were obscure. Even later, when Sykes et al. (1995) put forward the idea of the closed system, it was expected that the intracrystalline fraction should have contained highly conserved protein sequences, the primitive 'toolkit'.

Being able to find and analyse monospecific samples throughout a stratigraphic sequence is often difficult. However, when two or three genera can be analysed from the same site, this provides a powerful tool for checking the consistency of results (see discussion in Wehmiller et al., 2010). Furthermore, it may be possible to calculate 'conversion factors' in order to integrate D/L values obtained on different species, but this is risky as it relies on a perfect knowledge of the diagenetic trajectories of one or more species. Overall, the species effect offers insights into 'chemical taxonomy' and, also, the opportunity to develop multi-genera models able to link detailed data on amino acid sequences, protein structures, interaction with the mineral and the kinetics of degradation.

4.4.3   Burial Environment: Mixing

Localized mixing of the sediments may be reflected in age mixing, which is typical (but not exclusive) of high energy depositional environments, such as raised beaches or aeolian deposits (Wehmiller et al., 1995, 2015; Kosnik et al., 2007; Yanes et al., 2007; Krause et al., 2010; Kidwell, 2013; Torres et al., 2013). In fact, there are two different elements to this: the natural assemblage of shells that are found dead on a modern beach or in lagoonal sediments, for example, will reflect a range of ages, and sometimes hundreds or even thousands of years are represented (although there is a taxonomic effect to be taken into account, in that some species are more prone to surviving in death assemblages than others). The distribution of the ages of the constituents of an assemblage is called time averaging and represents an important factor in limiting the temporal resolution of AAR (Flessa and Kowalewski, 1994; Kowalewski et al., 1998). We discuss time averaging in this chapter with regard to its relevance for Holocene palaeoecology. The second, additional, element is the mixing of sediments of widely different ages: for example, a terrace of superimposed raised beaches each belonging to a high-sea level episode (commonly occurring during an interglacial or interstadial) means that the superimposed sediments might undergo a certain extent of mixing and reworking, upwards or downwards, and consequently that the shells within the sediments may date to the 'wrong' interglacial. This phenomenon of sediment recycling is visible also on Holocene timescales (e.g. Olszewski and Kaufman, 2015).

Reworking can be surprisingly helpful: for example, AAR dating of shells reworked by ice-sheet margins retreating and advancing can yield chronological information on fluctuations of glaciers during the Holocene (Briner et al., 2014). However, reworking has been invoked in (too) many cases in order to justify high natural variability of the D/L values, which would have limited age resolution. In 2002, McCarrol severely criticized the aminostratigraphc work which had thus far been carried out for the British Isles, arguing that the natural variability had been underestimated (McCarroll, 2002). The summary of his key critique is the starting point for a major part of the subsequent research in AAR: 'Aminostratigraphy is central to the recently revised correlation of Quaternary deposits in the British Isles, providing a link between terrestrial deposits and marine Oxygen Isotope Stages. [...] The data available suggest that amino acid ratios from different interglacials do not fall into discrete groups, but overlap considerably. It is therefore not valid to assign individual shells to Oxygen Isotope Stages simply on the basis of their amino acid ratios, which means that filtering data to remove high or low values, on the assumption that they represent reworked shells, is unacceptable.'

As a result of this critique and, in general, of the work carried out on the issue of reworking, two new areas of research in AAR have arisen since the late 2000s. The first aims at *reducing* natural variability by isolating the intracrystalline fraction. The second is dedicated to *estimating* the natural variability of D/L values, and to carrying out outlier analysis in a statistically significant way. Of great significance is the study by Kosnik and Kaufman (2008) which assessed a number of screening techniques able to identify outliers, i.e. samples with aberrant D/L values: the covariance between Asx and Glx D/L values (a lack of covariance indicates that the sample had been compromised during diagenesis), the concentration of Ser (high values of this labile amino acid suggest modern contamination), and the replication of measurements (to assess heterogeneity within the same specimen). They concluded that a combination of approaches is necessary, and especially highlighted the need to obtain data from multiple amino acids. To these two methods, we would add that a well-understood stratigraphic framework is fundamental for flagging up potentially compromised samples.

4.4.4   Burial Environment: Temperature

Thermal effects are crucial and cannot be ignored, both on a microscale and on a macroscale. Locally, samples might be exposed to varying temperatures (including diurnal and seasonal cycles) due to their burial environment (shallow or deep burial; sandy, silty or gravelly sediments, with different heat diffusion coefficients; sea floor, cave or open-air site; presence of decaying organic matter generating heat). This can result in differences in the extent of diagenesis even among samples from the same site or from the same thermal region (i.e. an area which today shows climatic homogeneity). The key issue is the relative length of time spent at higher than normal temperatures: the same time span of 'heating' will be almost irrelevant for a very old (Quaternary) sample but will have a great effect on a younger (e.g. Holocene) specimen. For example, Wehmiller (1977) could explain the variable A/I values obtained from shell samples in the Del Mar midden site (California) in terms of temperature fluctuations in the soil, using ground temperature data from the region. To minimize this effect, the practitioner will avoid sampling from the surface of an outcrop, but will target instead samples that are buried under at least 1 m of sediment (or at a depth where the seasonal temperature variations are dampened to less than 6 °C: Wehmiller and Miller, 2000). The threshold of 1 m derives from the fact that, at that depth, the surface temperatures (which are higher than the air temperature) decline and their fluctuations follow those of the mean annual temperatures (MAT) for the area. MATs are known and therefore the effect of MAT on the kinetics of racemization can be modelled (Wehmiller, 1977; Wehmiller et al., 2000). Recently, the effect of thermal age on the degradation of DNA and on the extent of AAR has been modelled by Collins et al., who set up an online tool for estimating the thermal age of a sample based on its burial and storage history (www.thermal-age.eu; see also Demarchi et al., 2016).

Although difficult to account for, there are nevertheless other factors affecting the local thermal history of a sample, the foremost of which concerns heat released as the result of cooking or other anthropogenic practices. This is especially important with regard to shell midden (dump) sites, which consist of accumulations of discarded mollusc shells found along the world's coasts and fluvial systems. These shell midden sites, which are often very large, were not only host to a range of human activities, such as cooking, but also held ceremonial/ritual significance. Hearths, for example, are an extremely common feature of these deposits, and, while this provides a precious source of charcoal for radiocarbon dating, it also provides a source of grief for the AAR scientist, because in the absence of distinctive blackening it is effectively impossible to tell by the naked eye if a sample has been exposed to heat for a short time or if, for example, it has been boiled. Indeed, the effect of short-time heating on the rate of diagenesis of proteins in biominerals is far from clear. Molluscs and bird eggs certainly represented an important food staple in the past, and some work aimed at identifying heating, burning or cooking on shells and eggshells has been carried out

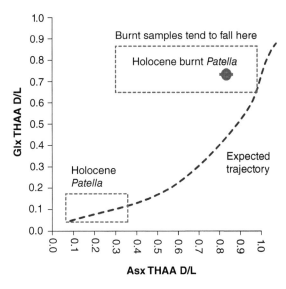

Figure 4.5 One possible way to identify burnt or heated samples using the co-variance of the D/L values of two amino acids. Schematic representation of data from Demarchi et al. (2011).

using a combination of methods. These include microscopic or spectroscopic techniques, which are able to identify crystal changes or phase transformation and recrystallization (e.g. from aragonite to calcite), geochemical methods (e.g. 'clumped isotopes', extremely useful for palaeotemperature reconstruction) and AAR (Brooks et al., 1991; Demarchi et al., 2011; Crisp, 2013; Miller et al., 2016; Müller et al., 2017). AAR-based 'burning indicators' include:

- presence of hydrophobic compounds (non-amino acids) in the chromatograms (late eluting);
- changes in the overall amino acid composition (especially due to Asp and Glu decarboxylation);
- deviation from the expected diagenesis pattern when the co-variance of the D/L values and other diagenesis indicators is considered (Figure 4.5).

4.5 Aminostratigraphy

As we have just summarized, AAR estimates can be affected by a range of factors which enhance natural variability at a microscale. However, at a macroscale, these effects can be considered negligible, and AAR can be used for cross-correlating stratigraphies across vast areas, using the approach of 'aminostratigraphy'. The analysis of multiple amino acids (and, where possible, multiple taxa: see Wehmiller et al., 2010, for example) is a crucial improvement of aminostratigraphy, since it exploits a natural property of the molecules themselves (they do not all racemize at the same rate) in order to build robust chrono-stratigraphic frameworks.

Amino acid racemization rates vary according to the position of the amino acid (terminal or internal) versus free amino acids, as we have discussed in Chapter 2. However, even amongst free amino acids there are striking differences in rate of racemization, which depend primarily on the side chain of the amino acid, i.e. whether its chemistry and steric characteristics hinder or aid proton abstraction. This is important because, if multiple D/L values can be measured, isochronous information becomes available. This fact has been exploited by Whiteacre et al. (2017), for example, when developing conversion factors. Furthermore, if we consider the THAA and FAA pools, we observe that they too yield different values, which are nevertheless related: the covariance between THAA and FAA of the same amino acid and the THAA or FAA of different amino acids provides a powerful method for building relative AAR chronologies, Figure 4.6.

It is important to highlight that the idea of using D/L values from the FAA and THAA fractions as a tool for building aminostratigraphies is relatively novel, having been fully exploited only after the concept of 'intracrystalline

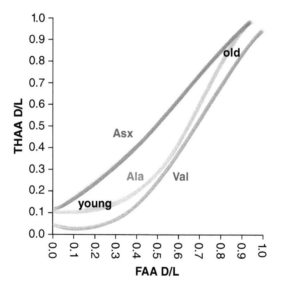

Figure 4.6 Hypothetical graph showing the extent of FAA and THAA racemization of a fast-racemizing (Asx), a medium-racemizing (Ala) and a slow-racemizing (Val) amino acid (trajectories based on D/L values measured on bleached *Patella* shells heated at 140 °C). Younger and less degraded samples tend to fall near the origin of the axes, while progressively older (more degraded) samples fall on the trajectory lines until they eventually reach equilibrium (top right). This plot enables the use of multiple independent measures of D/L values so as to build relative aminochronologies (e.g. to reconstruct the relative age of shells in Figure 4.3C).

protein dating' (IcPD) was introduced in the mid-1990s. However, aminostratigraphies had been successfully built as early as the 1960s. Miller, Hare and colleagues (Miller et al., 1979; Miller, 1980) employed the term 'aminostratigraphy' to describe the use of amino acid ratios (in molluscs) as a regional correlating tool. The adjective *regional* is significant here, since samples of the same age will yield different values if one has spent its burial history in colder climates than the other (the rate of reaction approximately doubles every 10 °C).

One of the most fruitful areas of research for aminostratigraphy has been its application to deep-sea sediments, since the constant temperatures found at the bottom of the ocean mean that the complicating effect of the interglacial–glacial cycles are dampened (Kvenvolden et al., 1973; Wehmiller, 1980). Fluvial terraces (whose stratigraphy can be related to their age in terms of Marine Isotope Stage record) have provided an excellent testing ground and independent dating for AAR on terrestrial settings (Bates, 1993; Penkman et al., 2007, 2011; Wehmiller, 2015b). Coastal sites (Figure 4.7) have been similarly favoured, not only because their stratigraphy is an indication of their relative ages (e.g. in the case of raised beach deposits) and can thus be used to test the coherence of D/L values, but also because their dating provides crucial information on sea levels during the regressions and transgressions that characterized the glacial–interglacial cycles. The coastal areas of North America comprise one of the main regions in which AAR has been developed since the 1980s; in his recent review of several important studies, Wehmiller (2013a) summarized the two great achievements of these first

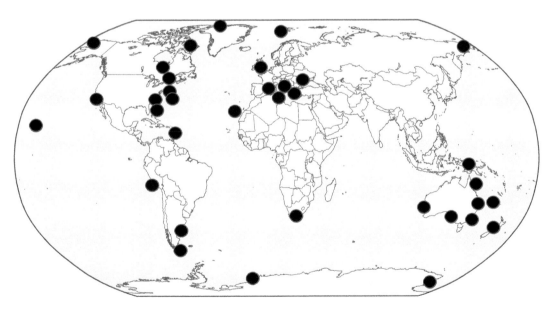

Figure 4.7 Approximate location of AAR 'coastal' studies, redrawn and simplified from Wehmiller (2015a). The reader interested in gaining an in-depth understanding of the application of AAR geochronology to coastal research should refer to that publication (Wehmiller, 2015a).

decades of aminostratigraphy on the Atlantic and Pacific coasts of North America thus:

1 in both uplifted marine terrace and subsurface sections, it can be demonstrated that D/L values increase with increasing geologic age;
2 D/L values in samples of equal age broadly increase with decreasing latitude (increasing temperature), although 'offsets and irregularities can be observed, which are likely caused by variations in the thermal gradient' (J.F. Wehmiller, personal communication, February 2019).

Geological and archaeological sites in the British Isles and the Mediterranean, as well as in Oceania and South Africa, have also been the subjects of in-depth aminostratigraphic (and aminochronological) studies (e.g. Bowen et al., 1985, 1989; Hearty et al., 1986; Brooks et al., 1990; Miller et al., 2000). The data accumulated over more than 40 years of research have been published in hundreds of scientific articles, and most of the raw data has been collected in the NOAA (National Oceanic and Atmospheric Administration) repository by many scientists involved in AAR research: ftp.ncdc.noaa.gov/pub/data/paleo/aar/. In particular, John Wehmiller has been at the forefront of making as many AAR datasets as possible publicly available. Examples of resources are the ArcGIS AAR and the University of Delaware AAR Database (UDAARDB) (Wehmiller and Pellerito, 2015).

The most recent comprehensive aminostratigraphic framework in Europe is the one built by Penkman and co-workers for the British Isles: they exploited the excellent closed-system behaviour of a special biomineral, i.e. the calcitic opercula of *Bithynia*, widespread in sediments and able to preserve original amino acids for over 30 million years (Penkman et al., 2013). Analysis of almost 500 opercula from 100 stratigraphic horizons resulted in a very robust framework, able to correctly assign samples to the interglacials of the isotope stage record back to the Early Pleistocene (Figure 4.8). The main limitation of this approach is that each MIS interglacial 'runs into' the next one, because in Britain the temperatures experienced by samples during glacial stages were so low as to practically halt racemization: the D/L values of the oldest part of one interglacial will therefore be similar to those of the youngest part of the preceding interglacial. Nonetheless, this framework has been successfully employed in order to solve a number of outstanding climatic, biostratigraphical and archaeological questions in the Quaternary of the British Isles (Penkman et al., 2011, 2013). Examples are: the existence of four post-MIS 12 interglacials within the Thames terrace sequence, supporting the *Mimomys savini/Arvicola* biostratigraphic model; the presence of the snail *Corbicula* (a warm-climate indicator) during the latter part of MIS 11, in MIS 9 and 7 but not MIS 5, while *Hippopotamus* was confirmed as present in the British Isles solely during the Last Interglacial (MIS 5e); the pre-MIS 12 age of the earliest human record in Britain; the presence of a range of industries, including Levallois, between MIS 11 and 7,

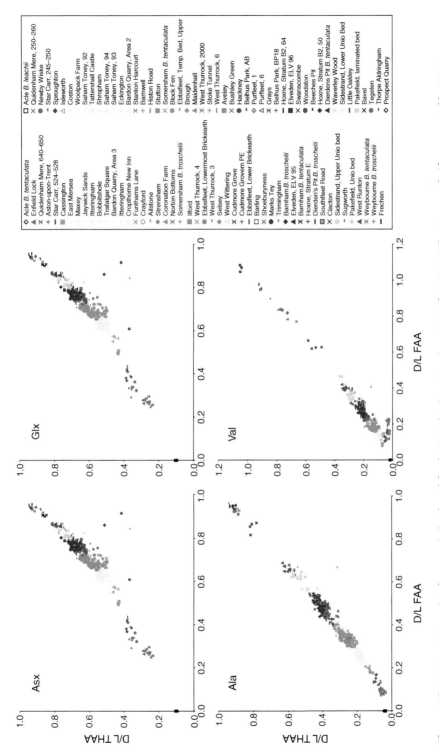

Figure 4.8 Aminostratigraphic framework developed for the British Isles using the extent of AAR in *Bithynia opercula* (figure created by Kirsty Penkman and reproduced with permission).

and the absence of humans during the Last interglacial (the 'deserted Britain' model).

4.6   Aminochronology

Aminochronology as defined by Wehmiller (1993) is a *calibrated aminostratigraphy* used as a regional chronometric technique. The relative aminostratigraphic frameworks can be calibrated if the D/L values are measured on samples of known age – for example, a series of shells that have been dated by radiocarbon (Kosnik et al., 2008). The resulting calibrated chronological frameworks can then be used to date samples independently, using equations describing the relationship between D/L values and time at a certain temperature. Three main approaches have been used.

3   Calibration based on kinetic models: this approach uses high-temperature experiments, in which samples are heated at known temperatures (usually between 80 and 140 °C), and the time–D/L relationship (the observed rate of reaction, k_{obs}) derived (see a review in Clark and Murray-Wallace, 2006). Usually, first-order reversible kinetics are assumed and an Arrhenius plot used to calculate the rate at high temperatures. This rate can then be used to extrapolate the age of samples at a certain burial temperature. However, data on a range of fossil biominerals and kinetic experiments have shown that, both on 'open-system' and 'closed-system' biominerals, the patterns of diagenesis at high and low temperatures do not follow the same trajectory (e.g. Wehmiller and Hare, 1971; Wehmiller and Belknap, 1978; Goodfriend and Meyer, 1991; Demarchi et al., 2013b; Tomiak et al., 2013) and therefore extrapolation to low temperatures is fraught with uncertainties.

4   Combination of high-temperature models and fossil samples of known age: this approach is safer, because it does not assume that the kinetics and pathways of reaction observed at high temperature follow those at low temperature and the Arrhenius plot is thus constrained at the burial temperature (Figure 4.9). However, it still implies a mechanistic understanding of the reaction, which we do not have.

5   Calibration using independently dated fossil samples: no prior knowledge of the reaction rate is assumed, and the time–AAR relationship is derived from the pairing of D/L values and numerical ages. The latter are typically obtained by U-series, OSL dating or Sr isotope geochronology for the Quaternary (see Umhoefer et al., 2014, and Oakley et al., 2017, for recent references). However, many successful applications of AAR geochronology have recently been in the field of Holocene (0–10 ka) research, because ^{14}C dating is well within

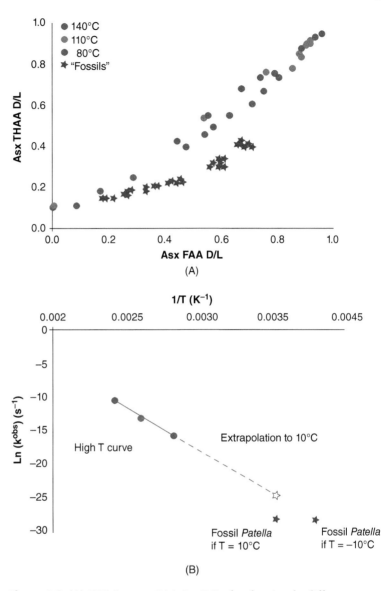

Figure 4.9 (A) THAA versus FAA Asx D/L plot showing the difference between the patterns of diagenesis in bleached *Patella* powders heated at high temperature (140 °C, 110 °C, 80 °C) and unheated but 'aged' naturally (i.e. archaeological and geological samples, with ages ranging from 500 to 250 000 years ago). (B) Arrhenius plot showing the results of extrapolating high temperature rate constants to low temperatures: the extrapolated 'fossil *Patella*' point (white star) is far from the real values (i.e. those obtained by calculating the reaction rate on fossil samples of known age for a hypothetical temperature of 10 °C or –10 °C, represented by the two blue stars in the figure). All data from Demarchi (2010).

its temporal comfort zone (0–50 ka) and because an even thermal history can be assumed for the samples, which have not been subjected to temperature variations due to the succession of glacial and interglacial cycles.

Given the lack of conformity to first-order kinetics (FOK) which implies a linear relationship between ln [(1 + D/L)/(1 – D/L)] and time, the modelling of AAR ages has typically been carried out using a variety of mathematical approaches: 'apparent parabolic kinetics' (APK), 'simple power-law kinetics' (SPK), and 'constrained power-law kinetics' (CPK) (e.g. Goodfriend, 1991; Kaufman, 2006; Wehmiller et al., 2012). These all use mathematical transformations in an attempt to linearize the D/L versus time relationship. In an effort to move on from these models, a 'model-free' approach has also been proposed, which uses successive optimizations (by 'scaling' of the time axis) of a curve to real D/L and %FAA data (high-temperature and fossil) in order to estimate Arrhenius parameters (Demarchi et al., 2013b; Tomiak et al., 2013). Such an approach, however, which effectively attempts to approximate the entire range of AAR values, shows that mimicking the complex network of reactions driving racemization (or hydrolysis) is extremely difficult. Another possibility is that of 'proportional time' (see Wehmiller et al., 2010, for example), which assumes that one can estimate age *differences* using the proportional changes in Asp D/L and Glu D/L, based on kinetic data from the heating experiments performed by Kaufman on foraminifera (Kaufman, 2006). Allen et al. (2013) have instead proposed a quasi-mechanistic model, which is grounded in FOK but allows the rate to decrease in time (thus better describing the way in which the D/L values asymptotically approach unity), which they called 'time-dependent kinetic' (TDK). The same authors also provided a much-needed tool (based on Bayesian statistics) to evaluate the fitting of the different calibration models and to estimate the uncertainties associated with age prediction, which are not (as usually assumed), normally distributed – in other words, older ages tend to be associated with larger uncertainties. This is certainly a step forward towards unifying the mathematical and chemical sides of AAR, but, as the authors point out, although TDK performs well in the majority of cases, in others it is no better (and in fact slightly worse) than other models. Until we reach a mechanistic understanding of the reactions occurring in the system, this impasse may not be broken.

Attempts have been made in this respect, using synthetic peptides and purified proteins degraded in high-temperature experiments in order to track diagenesis using a combination of AAR and mass spectrometry (MALDI-MS). However, they have generally met with limited success, due to the nonquantitative nature of the MALDI-MS data (Demarchi et al., 2013a). However, recent datasets (Demarchi et al., 2016), obtained on purified ostrich eggshell proteins heated at various times and temperatures and analysed by tandem mass spectrometry (LC-MS/MS), have shown that it may be feasible to track the hydrolytic sites of protein sequences and assess the frequency of breakage at various sites. It should therefore be possible to write a theoretical kinetic model that predicts the probability of a racemization event at each site. Obviously, this would be limited to those (extremely few) systems where the protein sequences of the whole system (and their structures) are known.

4.6.1 AAR and Holocene Palaeoecology

In the context of using aminochronology as a dating tool, the recent, and yet rapidly developing, field of AAR applied to Holocene palaeoecology merits a special mention. The ability to reconstruct the sedimentary history of specific ecosystems is very important if we want to use palaeoenvironmental data to assess (and predict) the impact of human activity on the biosphere and geosphere. As such, AAR data yield high- and medium-resolution temporal sequences for Holocene stratigraphies, which have been exploited in time-averaging, palaeoecological and taphonomic studies on mollusc shells but also on echinoderms and corals (Barbour Wood et al., 2003, 2006; Kosnik et al., 2007, 2015, 2017; Dexter et al., 2014; Ritter et al., 2017; New et al., 2019). These are time- and resource-intensive enterprises but have the merit of elucidating the potential of different sedimentary settings to be used for palaeoecology. As Domínguez et al. (2016, p. 564) put it: 'Time-averaging has evolved from an unrecognized variable in palaeoecological analyses to a key concept in understanding the dynamics of sedimentary systems and the formation of fossil deposits.'

Of note is the work conducted by Kosnik et al. (2015) on lagoons within the endangered Great Barrier Reef in Australia. Here it is important to assess the impact of recent anthropogenic activities on the reef (i.e. since Western colonization), and therefore sediments analysed for this purpose should be in stratigraphic order and represent relatively short time periods. To understand the extent of time averaging in these environments, sediment cores are typically taken and, from thin slices within those cores, tens or hundreds of biomineral samples are selected and analysed for AAR. A selection of these is then also directly dated by radiocarbon, and sometimes independent dating of the sediments is also available (Kosnik et al., 2015). These studies all employ robust statistics and age models developed on paired AAR/^{14}C dates for multiple amino acids, and contain a clear assessment of the contribution of each amino acid and each model to the final age estimates. The end result is an estimate of the extent of time averaging which affects the depositional environment, as well as an assessment as to whether the layers are in the correct order (i.e. youngest to oldest, top to bottom). This is crucial information: there is a big difference between analysing the palaeoecological signal from a sedimentary unit that has been thoroughly mixed and that represents a span of a few hundred (or even thousand) years, and one that represents time slices amounting to just a few years.

Other recent examples include: the study of ecological changes in Sydney harbour's benthic communities since colonization (Dominguez et al., 2016); the discovery that, contrary to foraminifera, dead *Nautilus* (cephalopods) shells typically only survive for a few hundred years (Tomašových et al., 2016); the environmental impact (or lack thereof) of oil platforms, e.g. in the Persian (Arabian) Gulf (Albano et al., 2016); the reconstruction of the Adriatic Sea's ecosystems at a submillennial scale (Tomašových et al., 2016, 2017, 2018, 2019); and exploring the potential for non-molluscan biominerals for fine-resolution studies, for example echinoids (Kowalewski et al., 2017) or sand dollars (Kosnik et al., 2017). Overall, Holocene studies are one of the most rapidly growing applications of AAR dating.

4.6.2 Quaternary Aminochronologies

Historically, the majority of AAR studies have essentially focused on Quaternary deposits, which lie well beyond the radiocarbon range: in this case, calibration must rely on other independent chronologies, for example, uranium series, optically stimulated luminescence (OSL), strontium isotopes stratigraphy, palaeomagnetism, biostratigraphy, and sedimentary evidence (e.g. Wehmiller, 1982; Hearty et al., 1986; Wehmiller et al., 1988, 2012). Calibration is especially important because pre-Holocene samples have experienced a number of glacial–interglacial cycles and have therefore been exposed to different burial temperatures, which prevent Holocene-based calibrations from being extended back to the Pleistocene.

The most suitable setting for developing age models into the Quaternary are therefore deep-sea cores, in which sediments rich in foraminifera have accumulated continuously. Foraminifera tests (or bulk samples) can be assumed to have experienced the same diagenetic temperature, and the cores can also yield independent age estimates using a variety of proxies (including correlation of oxygen isotopes ratios with the MIS record). Therefore, D/L values can be calibrated in a relatively straightforward way and used to build aminochronological frameworks. This is the reason for which the first studies on racemization dating and kinetics were performed on marine sediments (Wehmiller and Hare, 1971; Kvenvolden et al., 1973; King and Neville, 1977; Bada et al., 1978; Müller, 1984).

Far more common is the matter of having to date subaerial stratigraphies (coastal, fluvial, terrestrial), which are usually discontinuous. Dating these exposures is important because it gives direct information on the extent of transgressions and regressions of sea levels during interglacial and glacial cycles, on the extent of ice cap cover, on the actual number and succession of cold and warm stages (and their impact on fauna and vegetation), and on the history and evolution of riverine systems. These sediments will therefore have experienced a succession of warm and cool stages: we have already seen how, in the British Isles, this has had the practical effect of making the transition between D/L values which belong to samples from two successive interglacials practically indistinguishable (Penkman et al., 2011). Classic examples are the framework based on calibrated aminozones (statistically significant clusters of D/L values) developed by Hearty and colleagues for Mediterranean raised beaches (Hearty et al., 1986) and the frameworks established for the Atlantic and Pacific coasts of North America by Wehmiller and colleagues (Wehmiller, 2013a). In both instances, a strong north–south temperature gradient can be observed, with D/L (A/I) values increasing not only with age, but also with temperature. The number of studies concerning the dating of Quaternary stratigraphies is practically uncountable, and considering the various applications is beyond the scope of this book. Reviews include: Schroeder and Bada (1976), Miller and Brigham-Grette (1989), Demarchi and Collins (2015) (geologists and archaeologists alike are encouraged to read these for a better insight into the evolution of this dating technique) and Wehmiller (2015a, 2015b). The main concern here is to give tips and pointers for setting up an AAR study, which are detailed in Section 4.10.

4.7   Palaeothermometry

The ability to back-calculate the effective temperature (i.e. the integrated thermal history) experienced by a sample on the basis of its numerical age and extent of degradation stems directly from the ability to date AAR samples both directly and independently and using these paired values to develop accurate kinetic models, which can link time to D/L values (rate of racemization). John Wehmiller has been one of the main scientists involved in the development of palaeothermometry since the 1970s (Wehmiller, 1977; Wehmiller and Belknap, 1978) with his study of the thermal history of the samples from Californian Del Mar shell middens.

Ideally, the paired values would be ^{14}C/AAR, but for Quaternary stratigraphies this is not possible beyond 40–50 ka BP (the time limit for radiocarbon). This was the case for one of the most famous applications of AAR palaeothermometry: the estimate of the extent of cooling (9 °C between the period 50–16 ka ago and 16 ka–present) in the Southern Hemisphere during the last glacial maximum using ^{14}C-dated emu eggshell. Emu eggshell yields quasi-linear apparent racemization kinetics, and this study was thus able to demonstrate that cooling in Australia was dramatic and began well before the Last Glacial Maximum, while warming also began well before the onset of the Holocene, and also that such phenomena were forced by global temperature changes (Miller et al., 1997). When direct ^{14}C dating is not possible, the sediments surrounding the sample can be dated by OSL (aeolian deposits especially) or U-Th (if coral or flowstone are present). In the Northern Hemisphere, Kaufman used AMS ^{14}C, luminescence, tephra and amino acid geochronology on ostracods in order to reconstruct the temperature history of the Bonneville Basin, Utah (Kaufman, 2003), while Oches and McCoy worked on the Mississippi Valley (Oches et al., 1996); McCoy also provided estimates of the precision of palaeothermometry (McCoy, 1987). An alternative direct dating method, applicable to shells, for example, is Sr isotopic geochronology (Wehmiller et al., 2012). These age estimates can be used to predict kinetic curves for samples which have experienced different effective temperatures (e.g. Wehmiller et al., 2012, their Figure 6).

Wehmiller et al. (2012) set out the general procedures for palaeothermometry as follows:

1 create a model temperature history (in 1000-year time intervals) based on a generalized version of the MIS, normalizing this history to the present temperature in the study area;
2 use this model temperature history to calculate the effective racemization rate constant for each time interval, using the experimentally derived temperature sensitivity of the rate constant k;
3 calculate the average rate constant over the entire history of a sample;
4 convert this average rate constant into an effective temperature for a sample of any selected age;
5 compare this modelled effective temperature with the effective temperature derived from the calibrated kinetic model.

A similar approach was also used to derive the effective temperature and to normalize the thermal age of samples of ostrich eggshell from African sites with different temperatures, expressing each thermal age in 'years at 10 °C' (Demarchi et al., 2016). Broadly speaking, converting effective temperatures to palaeotemperature histories requires knowledge of the timing and duration of glacial cycles, the extent of cooling and warming, and an accurate determination of temperature sensitivities. Palaeothermometry models are very elegant, but uncertainties in age estimates relating to both kinetic modelling and temperature histories lead to an overall error in palaeotemperature estimates. Nonetheless, palaeothermometry does have the potential of adding valuable data to palaeoclimatic reconstructions, and is an application of AAR which merits more attention than it has thus far received.

4.8 Testing the Suitability of Biominerals for Geochemical Analyses

This application of AAR is relatively recent and stems from the combined use of intracrystalline protein geochronology (i.e. the variant of AAR dating that includes the isolation of a closed system) and stable isotope geochemistry in order to reconstruct palaeoenvironmental conditions, especially for archaeological sites. Even for those cases for which the temporal scale involved is too short and the AAR resolution insufficient for geochronological applications, the co-variance of THAA and FAA D/L values (or, more generally, of two diagenetic proxies measured within the closed system) is useful for identifying samples which have been somehow diagenetically compromised, i.e. exhibit non-closed-system behaviour. Such samples might have been heated, for example, or the mineral might have recrystallized, e.g. aragonite might have turned to calcite. Both heating and recrystallization have an impact on the accuracy of palaeoenvironmental and palaeotemperature reconstructions using stable oxygen isotopes (Müller et al., 2017). Therefore, being able to detect these potential issues might be useful for evaluating the significance of other geochemical datasets. One such example is the analysis of *Phorcus lineatus* shells from the Early Upper Palaeolithic site of Ksâr 'Akil, Lebanon: AAR did not have sufficient resolution for dating, but was used to identify shells which, due to their being potentially compromised, could have yielded problematic radiocarbon or stable isotope data (Bosch et al., 2015a).

4.9   Taxonomic Identification

There is one further application of AAR analysis that has been attempted several times in the past: using the bulk amino acid composition as a way of obtaining taxonomic classification of biominerals from archaeological or geological sites. This application is especially useful for samples which are highly fragmented or heavily worked, as both these processes obliterate morphological details that are key for identification. For example, in order to produce tiny perforated beads that can be used for necklaces, armbands or other ornamental apparatus, it is necessary to cut the raw material (bone, shell, eggshell, ivory, antler, stone) into small pieces, which are then abraded until they take a circular shape, perforated using a drill, and finally polished. It is possible to use optical and electronic microscopy in order to gather information on the microstructure of the biomineral and a variety of spectroscopic techniques for clarifying the mineralogy and chemical composition of the material, but even the combination of these approaches is unlikely to be sufficient for obtaining a firm identification.

The idea of using the amino acid composition of fragmentary shells for their identification was originally put forward by Andrews et al. (1985) and further developed by Kaufman et al. (1992) and, for benthic foraminifera, by Haugen et al. (1989). Demarchi et al. (2014) proposed that the bulk amino acid (THAA) composition of bleached shell powders, i.e. the intracrystalline fraction, could be used as a taxonomic tool for the identification of small shell samples. In particular, they used the relative concentrations of six amino acids, normalized so that the sum of the six amino acid concentrations was the same value for each sample. They tested 777 shell samples belonging to 29 genera from 27 families and 15 orders and used the six variables for classification by learning vector quantization (LVQ) and for coefficient of similarity (CS) calculations. They found that ~77% classifications were correct at the level of genus and 84% at the level

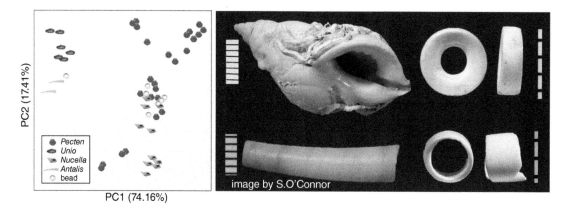

Figure 4.10 Principal component analysis (PCA) plot comparing the relative normalized amino acid composition of archaeological beads from the Early Bronze Age site of Great Cornard (UK) with some marine and freshwater shells. The study revealed that the most likely identification for the raw material used to manufacture these beads were the marine gastropod *Nucella* and the scaphopod *Antalis*. Original figures by Demarchi, Wilson and O'Connor, published in Demarchi et al. (2014), reproduced under the Creative Commons Attribution (CC BY).

of order. However, the dataset used was not ideal because some taxa were represented by a high number of samples, and some by only a few shells, thus potentially skewing the assessment of natural variability of the data. Furthermore, taxa that are not present in this database obviously cannot be taken into account as a potential species – it must be remembered that there are more than 100 000 molluscan species (both extant and extinct); therefore, hoping to build a comprehensive database is somewhat optimistic. Nonetheless, this approach can be useful for those instances in which the understanding of the relative composition of the molluscan fauna assemblage is important, for example, to track changes in subsistence strategies or environmental shifts in the past. The study by Demarchi et al. led to the discovery that locally sourced shells (*Nucella* and *Antalis*) were likely used to make an elaborate chest ornament found in a high-status Early Bronze Age woman's grave (Figure 4.10).

4.10 Appendix: Practical Tips on How to Plan and Conduct an AAR Study

As soon as one begins to think of incorporating AAR into an archaeological, geological or palaeontological study, one should contact one of the laboratories that are currently active around the world:

- Europe: (1) University of York, North-east Amino Acid racemization, UK (Kirsty Penkman); (2) Universidad Politécnica de Madrid, Spain (Trinidad Torres, José Eugenio Ortiz)

- USA: (1) Northern Arizona University (Darrell Kaufman), (2) INSTAAR, University of Colorado Boulder (Giff Miller)
- Australia: University of Wollongong (Colin Murray-Wallace).

They will be able to help with research design, from sampling to data analysis and dissemination, and can advise on the suitability of the substrate of choice for AAR analyses. The cost can vary depending on the number of samples and effort of the personnel involved, but it is typically a fraction of the cost of radiocarbon dating. Hosting and training researchers is one of the main missions of the AAR labs, and, as such, collaborations are usually possible.

4.10.1 Preliminary Tests

If the biomineral(s) chosen for the study have never been used for AAR geochronology, the practitioners will usually recommend conducting a few preliminary tests in the laboratory in order to assess suitability for geochronology. These typically involve:

- bleaching experiments, where modern (or fossil) powdered biomineral is soaked in NaOCl for increasing times, and the FAA and THAA concentrations measured – this test reveals whether the biomineral retains a fraction of intracrystalline proteins;
- leaching experiments, in which bleached powders are heated in water in sealed glass ampoules at high temperature for increasing times (typically between 3 and 5 time points at 140 °C), the concentration of amino acids leached in the water is measured and the patterns of diagenesis checked for internal consistency – this test shows whether the intracrystalline fraction behaves as a closed system (no leaching, predictable diagenesis).

4.10.2 Sampling Strategies in the Field

Once the tests have been successfully carried out, the sampling campaign in the field can begin. It is highly recommended that the AAR specialist be present so that they can actively select sampling location and size. Best practice is to collect samples from >1 m below the surface, therefore avoiding the kinetically active zone (Wehmiller et al., 2000). Monospecific samples should be targeted and it is recommended that multiple taxa from each sampling location/layer be collected; for each species, a minimum of three specimens is needed per layer, and at least one layer should be more densely sampled (more than seven specimens) in order to calculate the natural variability (i.e. the spread of the data around the mean, which impacts on age resolution).

All specimens should be stored in clean sample bags, taking care to blot them dry before storage if necessary, but avoiding exposure to heat (i.e. drying in ovens or in direct sunlight).

4.10.3   Sampling Strategies in the Laboratory

If the specimens are small (e.g. foraminifera), or comprise a simple or consistent mineralogy/microstructure (e.g. eggshell), then analysing one sample per specimen should be sufficient. However, in the case of biominerals with complex microstructures and mineralogies, such as mollusc shells, preliminary work should be aimed at testing the intraspecimen variability and at choosing a consistent sampling position, if possible targeting a calcitic layer in shells, and avoiding the less stable aragonite. Ontogenic effects might also be present and therefore it would be best to avoid measuring D/L values on juvenile versus adult specimens.

Fossil samples should be cleaned using a combination of sonication in ultrapure water, brushing of the surface (clean brushes) and abrasion by drilling (clean diamond tip, bleached between samples) in order to clean the surfaces from concretions or colonizing organisms (e.g. barnacles).

Dry, clean specimens should then be subsampled in the desired location, and any unwanted microstructural layers removed by drilling. Heat-inducing operations should be minimized since they can accelerate protein degradation locally. As such, wherever possible subsampling should be performed using mechanical devices such as tweezers and pliers, rather than collecting the powder generated by drilling.

The sample size required for AAR is small (10 mg is more than sufficient for several repeated analyses), therefore only a small fragment of biomineral is necessary. This will then be powdered using a clean mortar and pestle (cleaned by repeated rinses in 12% bleach, methanol and water), and the desired size fraction (typically <500 µm) selected by sieving.

4.10.4   Bleaching

Between 5 and 10 mg of powder will be accurately weighed out and transferred to a clean plastic Eppendorf tube, and 50 µL of 12% bleach added for each mg of powder. The results of the preliminary bleaching experiments will have provided information on the optimal bleaching time for that specific type of biomineral (usually 48 h for shell and 72 h for eggshell). Powders will be left to soak and shaken every 24 h in order to ensure penetration of the bleach. After the allotted time, bleach will be removed by rinsing the powders in ultrapure water and methanol (5 water washes and 1 methanol wash; each followed by centrifugation and removal of the supernatant). A bleach blank must be included in this step.

4.10.5 FAA and THAA

Bleached powders are split into subsamples (between 2 and 5 mg each, accurately weighed in glass vials, sterilized by baking at 450 °C for a minimum of 6 hours) for the analysis of FAA and THAA. FAA are released by demineralizing in cold weak hydrochloric acid (2 M HCl, a minimum of 10 µL per mg, usually 15–30 minutes), and THAA by acid hydrolysis at high temperatures (typical conditions are 7 M HCl, 24 h, 110 °C, 20 µL per mg). HCl blanks must be included in these steps. Both THAA and FAA are then evaporated until dry using a centrifugal evaporator (time required depends on the evaporator, but a minimum of 6–8 hours). Dried FAA and THAA can be stored at room temperature, waiting to be rehydrated (in a rehydration fluid, which contains a spike of the internal standard L-homo-arginine) for analysis.

4.10.6 Chromatography

The rehydration and subsequent analysis of samples and blanks (typically, by RP-HPLC) is performed by the AAR specialist, who will usually perform a double injection for each vial in order to assess analytical error. The analytical error will then be evaluated together with the other sources of uncertainty (intra-specimen variability, within-layer natural variability). Standard amino acid mixtures of known D/L value are interspersed throughout the analytical run and allow an assessment of instrument performance. Each laboratory also routinely analyses a set of inter-laboratory calibration standards (Powell et al., 2013; Wehmiller, 2013b), thus monitoring the performance of the instrument against that of all other laboratories worldwide. This is why it is important that AAR analysis be carried out in well-recognized facilities rather than using in-house instrumentation.

4.10.7 Data Analysis, Publication and Sharing

Data analysis will typically be carried out by the AAR specialists, who will derive D/L values and concentrations from chromatograms and perform outlier analysis and any modelling of the D/L values as appropriate. It is very important that the AAR specialist be involved in the evaluation of the results for publication, else there is a high risk of over-interpretation by the nonspecialist. AAR results will be included as raw data (D/L and concentrations as a minimum) in the Supplementary Information of all publications reporting on the use of AAR geochronology. Data will also be stored online in the appropriate repository: ftp.ncdc.noaa.gov/pub/data/paleo/aar/.

References

Albano, P.G., Filippova, N., Steger, J. et al. (2016). Oil platforms in the Persian (Arabian) Gulf: Living and death assemblages reveal no effects. *Cont. Shelf Res.*, 121: 21–34.

Allen, A.P., Kosnik, M.A., and Kaufman, D.S. (2013). Characterizing the dynamics of amino acid racemization using time-dependent reaction kinetics: A Bayesian approach to fitting age-calibration models. *Quat. Geochronol.*, 18: 63–77.

Andrews, J.T., Miller, G.H., Davies, D.C., and Davies, K.H. (1985). Generic identification of fragmentary Quaternary molluscs by amino acid chromatography: A tool for Quaternary and palaeontological research. *Geol. J.*, 20: 1–20.

Bada, J.L. (1985). Aspartic acid racemization ages of California Paleoindian skeletons. *Am. Antiq.*, 50: 645–647.

Bada, J.L. and Protsch, R. (1973). Racemization reaction of aspartic acid and its use in dating fossil bones. *Proc. Natl. Acad. Sci. U.S.A.*, 70: 1331–1334.

Bada, J.L., Schroeder, R.A., and Carter, G.F. (1974). New evidence for the antiquity of man in North America deduced from aspartic acid racemization. *Science*, 184: 791–793.

Bada, J.L., Shou, M.-Y., Man, E.H., and Schroeder, R.A. (1978). Decomposition of hydroxy amino acids in foraminiferal tests: Kinetics, mechanism and geochronological implications. *Earth Planet. Sci. Lett.*, 41: 67–76.

Bailey, G. (2004). World prehistory from the margins: The role of coastlines in human evolution. *J. Interdiscip. Stud. Hist. Archaeol.*, 11.

Barbour Wood, S.L., Krause, R.A., Jr, Kowalewski, M. et al. (2003). A comparison of rates of time averaging between the bivalve *Macoma cleryana* and brachiopod *Bouchardia rosea* on a shallow subtropical shelf. In: *Geol. Soc. Am., Abstracts with Programs*, 35, 273.

Barbour Wood, S.L., Krause, R.A., Kowalewski, M. et al. (2006). Aspartic acid racemization dating of Holocene brachiopods and bivalves from the southern Brazilian shelf, South Atlantic. *Quat. Res.*, 66: 323–331.

Bard, E., Hamelin, B., and Fairbanks, R.G. (1990). U-Th ages obtained by mass spectrometry in corals from Barbados: sea level during the past 130,000 years. *Nature*, 346: 456–458.

Bates, M.R. (1993). Quaternary aminostratigraphy in Northwestern France. *Quat. Sci. Rev.*, 12: 793–809.

Bender, M.L., Fairbanks, R.G., Taylor, F.W. et al. (1979). Uranium-series dating of the Pleistocene reef tracts of Barbados, West Indies. *Geol. Soc. Am. Bull.*, 90: 1577–1594.

Bosch, M.D., Mannino, M.A., Prendergast, A.L. et al. (2015a). New chronology for Ksâr 'Akil (Lebanon) supports Levantine route of modern human dispersal into Europe. *Proc. Natl. Acad. Sci. U.S.A.*, 112: 7683–7688.

Bosch, M.D., Mannino, M.A., Prendergast, A.L. et al. (2015b). Reply to Douka et al.: Critical evaluation of the Ksâr 'Akil chronologies. *Proc. Natl. Acad. Sci. U.S.A.* https://doi.org/10.1073/pnas.1520412112.

Bowen, D.Q., Sykes, G.A., Reeves (nee Henry), A. et al. (1985). Amino acid geochronology of raised beaches in south west Britain. *Quat. Sci. Rev.*, 4: 279–318.

Bowen, D.Q., Hughes, S., Sykes, G.A., and Miller, G.H. (1989). Land–sea correlations in the Pleistocene based on isoleucine epimerization in non-marine molluscs. *Nature*, 340: 49–51.

Briant, R.M., Brock, F., Demarchi, B. et al. (2018). Improving chronological control for environmental sequences from the last glacial period. *Quat. Geochronol.*, 43: 40–49.

Briner, J.P., Kaufman, D.S., Bennike, O., and Kosnik, M.A. (2014). Amino acid ratios in reworked marine bivalve shells constrain Greenland Ice Sheet history during the Holocene. *Geology*, 42: 75–78.

Brooks, A.S., Hare, P.E., Kokis, J.E. et al. (1990). Dating Pleistocene archeological sites by protein diagenesis in ostrich eggshell. *Science,* 248: 60–64.

Brooks, A.S., Hare, P.E., Kokis, J.E., and Durana, K. (1991). A burning question: Differences in laboratory induced and natural diagenesis in ostrich eggshell proteins. In: *Annu. Rep. Geophys. Lab., Carnegie Institution of Washington*, 176–179.

Campisano, C. (2012). Milankovitch cycles, paleoclimatic change, and hominin evolution. *Nature Ed. Knowl.*, 4: 5.

Chappell, J. and Shackleton, N.J. (1986). Oxygen isotopes and sea level. *Nature*, 324: 137–140.

Clark, S.J. and Murray-Wallace, C.V. (2006). Mathematical expressions used in amino acid racemization geochronology: A review. *Quat. Geochronol.*, 1: 261–278.

Collins, M.J., Waite, E.R., and van Duin, A.C. (1999). Predicting protein decomposition: The case of aspartic acid racemization kinetics. *Philos. Trans. R. Soc. Lond. B Biol. Sci.*, 354: 51–64.

Collins, M.J., Penkman, K.E.H., Rohland, N. et al. (2009). Is amino acid racemization a useful tool for screening for ancient DNA in bone? *Proc. Biol. Sci.*, 276: 2971–2977.

Crisp, M., Demarchi, B., Collins, M. et al. (2013). Isolation of the intra-crystalline proteins and kinetic studies in *Struthio camelus* (ostrich) eggshell for amino acid geochronology. *Quat. Geochronol.*, 16: 110–128.

Crisp, M.K. (2013). Amino acid racemization dating: Method development using African ostrich (*Struthio camelus*) eggshell. Unpublished PhD thesis, University of York.

Demarchi, B. (2010). Geochronology of coastal prehistoric environments: A new closed system approach using amino acid racemisation. Unpublished PhD thesis, University of York.

Demarchi, B. and Collins, M. (2015). Amino acid racemization dating. In: *Encyclopedia of Scientific Dating Methods*, 13-26. Springer.

Demarchi, B., Williams, M.G., Milner, N. et al. (2011). Amino acid racemization dating of marine shells: A mound of possibilities. *Quat. Int.*, 239: 114–124.

Demarchi, B., Collins, M., Bergström, E. et al. (2013a). New experimental evidence for in-chain amino acid racemization of serine in a model peptide. *Anal. Chem.*, 85: 5835–5842.

Demarchi, B., Collins, M.J., Tomiak, P.J. et al. (2013b). Intra-crystalline protein diagenesis (IcPD) in *Patella vulgata*. Part II: Breakdown and temperature sensitivity. *Quat. Geochronol.*, 16: 158–172.

Demarchi, B., O'Connor, S., de Lima Ponzoni, A. et al. (2014). An integrated approach to the taxonomic identification of prehistoric shell ornaments. *PLoS One*, 9: e99839.

Demarchi, B., Hall, S., Roncal-Herrero, T. et al. (2016). Protein sequences bound to mineral surfaces persist into deep time. *eLife,* 5: e17092.

Dexter, T.A., Kaufman, D.S., Krause, R.A. et al. (2014). A continuous multi-millennial record of surficial bivalve mollusk shells from the São Paulo Bight, Brazilian shelf. *Quat. Res.*, 81: 274–283.

Dincauze, D.F. (1984). An archaeological evaluation of the case for pre-Clovis occupations. *Adv. World Archaeol.*, 3: 275–323.

Dominguez, J.G., Kosnik, M.A., Allen, A.P. et al. (2016). Time-averaging and stratigraphic resolution in death assemblages and Holocene deposits: Sydney Harbour's molluscan record. *Palaios*, 31: 563–574.

Douka, K., Bergman, C.A., Hedges, R.E.M. et al. (2013). Chronology of Ksar Akil (Lebanon) and implications for the colonization of Europe by anatomically modern humans. *PLoS One*, 8: e72931.

Douka, K., Higham, T.F.G., and Bergman, C.A. (2015). Statistical and archaeological errors invalidate the proposed chronology for the site of Ksar Akil. *Proc. Natl. Acad. Sci. U.S.A.*, 112: E7034.

Emiliani, C. (1954). Depth habitats of some species of pelagic Foraminifera as indicated by oxygen isotope ratios. *Am. J. Sci.*, 252: 149–158.

Fitznar, H.P., Lobbes, J.M., and Kattner, G. (1999). Determination of enantiomeric amino acids with high-performance liquid chromatography and pre-column derivatisation with *o*-phthaldialdehyde and *N*-isobutyrylcysteine in seawater and fossil samples (mollusks). *J. Chromatogr. A*, 832: 123–132.

Fleming, K., Johnston, P., Zwartz, D. et al. (1998). Refining the eustatic sea-level curve since the Last Glacial Maximum using far- and intermediate-field sites. *Earth Planet. Sci. Lett.*, 163: 327–342.

Flessa, K.W. and Kowalewski, M. (1994). Shell survival and time-averaging in nearshore and shelf environments: Estimates from the radiocarbon literature. *Lethaia*, 27: 153–167.

Goodfriend, G.A. (1991). Patterns of racemization and epimerization of amino acids in land snail shells over the course of the Holocene. *Geochim. Cosmochim. Acta*, 55: 293–302.

Goodfriend, G.A. and Meyer, V.R. (1991). A comparative study of the kinetics of amino acid racemization/epimerization in fossil and modern mollusk shells. *Geochim. Cosmochim. Acta*, 55: 3355–3367.

Hare, E. and Abelson, P.H. (1968). Racemization of amino acids in fossil shells. *Year B. Carnegie Inst. Wash.*, 66: 526–528.

Hare, P.E. and Mitterer, R.M. (1969). Laboratory simulation of amino acid diagenesis in fossils. *Carnegie Institution of Washington Yearbook*, 67: 205–208.

Hare, P.E., Hoering, T.C., and King, K. (eds.) (1978). *Biogeochemistry of Amino Acids*. New York: Wiley.

Haslett, S.K. (2009). *Coastal Systems*. Routledge.

Haugen, J.-E., Sejrup, H.P., and Vogt, N.B. (1989). Chemotaxonomy of Quaternary benthic foraminifera using amino acids. *J. Foraminiferal Res.*, 19: 38–51.

Hearty, P.J., Miller, G.H., Stearns, C.E., and Szabo, B.J. (1986). Aminostratigraphy of Quaternary shorelines in the Mediterranean basin. *Geol. Soc. Am. Bull.*, 97: 850–858.

Hearty, P.J. and Kaufman, D.S. (2000). Whole-rock aminostratigraphy and quaternary sea-level history of the Bahamas. *Quat. Res.*, 54: 163–173.

Higham, T., Douka, K., Wood, R. et al. (2014). The timing and spatiotemporal patterning of Neanderthal disappearance. *Nature*, 512: 306–309.

Higham, T.F.G., Jacobi, R.M., and Bronk Ramsey, C. (2006). AMS radiocarbon dating of ancient bone using ultrafiltration. *Radiocarbon*, 48: 179–195.

Hill, R.L. (1965). Hydrolysis of proteins. *Adv. Protein Chem.*, 20: 37–107.

Hillaire-Marcel, C., Carro, O., Causse, C. et al. (1986). Th/U dating of *Strombus bubonius*-bearing marine terraces in southeastern Spain. *Geology*, 14: 613–616.

Huntley, D.J., Godfrey-Smith, D.I., and Thewalt, M.L.W. (1985). Optical dating of sediments. *Nature*, 313: 105–107.

Kaufman, D.S. (2003). Amino acid paleothermometry of Quaternary ostracodes from the Bonneville Basin, Utah. *Quat. Sci. Rev.*, 22: 899–914.

Kaufman, D.S. (2006). Temperature sensitivity of aspartic and glutamic acid racemization in the foraminifera Pulleniatina. *Quat. Geochronol.*, 1: 188–207.

Kaufman, D.S. and Manley, W.F. (1998). A new procedure for determining dl amino acid ratios in fossils using reverse phase liquid chromatography. *Quat. Sci. Rev.*, 17: 987–1000.

Kaufman, D.S., Miller, G.H., and Andrews, J.T. (1992). Amino acid composition as a taxonomic tool for molluscan fossils: An example from Pliocene–Pleistocene Arctic marine deposits. *Geochim. Cosmochim. Acta*, 56: 2445–2453.

Kidwell, S.M. (2013). Time-averaging and fidelity of modern death assemblages: Building a taphonomic foundation for conservation palaeobiology. *Palaeontology*, 56: 487–522.

King, K., Jr. and Neville, C. (1977). Isoleucine epimerization for dating marine sediments: Importance of analyzing monospecific foraminiferal samples. *Science*, 195: 1333–1335.

Kosnik, M.A. and Kaufman, D.S. (2008). Identifying outliers and assessing the accuracy of amino acid racemization measurements for geochronology. II: Data screening. *Quat. Geochronol.*, 3: 328–341.

Kosnik, M.A., Hua, Q., Jacobsen, G.E. et al. (2007). Sediment mixing and stratigraphic disorder revealed by the age-structure of *Tellina* shells in Great Barrier Reef sediment. *Geology*, 35: 811–814.

Kosnik, M.A., Kaufman, D.S., and Hua, Q. (2008). Identifying outliers and assessing the accuracy of amino acid racemization measurements for geochronology. I: Age calibration curves. *Quat. Geochronol.*, 3: 308–327.

Kosnik, M.A., Hua, Q., Kaufman, D.S., and Zawadzki, A. (2015). Sediment accumulation, stratigraphic order, and the extent of time-averaging in lagoonal sediments: A comparison of ^{210}Pb and ^{14}C/amino acid racemization chronologies. *Coral Reefs*, 34: 215–229.

Kosnik, M.A., Hua, Q., Kaufman, D.S. et al. (2017). Radiocarbon-calibrated amino acid racemization ages from Holocene sand dollars (*Peronella peronii*). *Quat. Geochronol.*, 39: 174–188.

Kowalewski, M., Goodfriend, G.A., and Flessa, K.W. (1998). High-resolution estimates of temporal mixing within shell beds: The evils and virtues of time-averaging. *Paleobiology*, 24: 287–304.

Kowalewski, M., Casebolt, S., Hua, Q. et al. (2017). One fossil record, multiple time resolutions: Disparate time-averaging of echinoids and mollusks on a Holocene carbonate platform. *Geology*, 46: 51–54.

Krause, R.A., Barbour, S.L., Kowalewski, M. et al. (2010). Quantitative comparisons and models of time-averaging in bivalve and brachiopod shell accumulations. *Paleobiology*, 36: 428–452.

Kvenvolden, K.A., Peterson, E., Wehmiller, J., and Hare, P.E. (1973). Racemization of amino acids in marine sediments determined by gas chromatography. *Geochim. Cosmochim. Acta*, 37: 2215–2225.

Lajoie, K.R., Wehmiller, J.F., and Kennedy, G.L. (1980). Inter-and intrageneric trends in apparent racemization kinetics of amino acids in Quaternary mollusks. In: *Biogeochemistry of Amino Acids* (eds. P.E. Hare, T.C. Hoering and K. King Jr.), 305–340. New York: Wiley.

Lambeck, K. and Chappell, J. (2001). Sea level change through the last glacial cycle. *Science*, 292: 679–686.

Li, W.X., Lundberg, J., Dickin, A.P. et al. (1989). High precision mass-spectrometric uranium-series dating of cave deposits and implications for palaeoclimate studies. *Nature*, 339: 534–536.

Lowe, J. and Walker, M.J.C. (2014). *Reconstructing Quaternary Environments*. Routledge.

Mcbrearty, S. and Brooks, A.S. (2000). The revolution that wasn't: A new interpretation of the origin of modern human behavior. *J. Hum. Evol.*, 39: 453–563.

McCarroll, D. (2002). Amino acid geochronology and the British Pleistocene: Secure stratigraphical framework or a case of circular reasoning? *J. Quat. Sci.*, 17: 647–651.

McCoy, W.D. (1987). The precision of amino acid geochronology and paleothermometry. *Quat. Sci. Rev.*, 6: 43–54.

Millard, A.R. (2008). A critique of the chronometric evidence for hominid fossils. I: Africa and the Near East 500–50 ka. *J. Hum. Evol.*, 54: 848–874.

Miller, G.H. and Brigham-Grette, J. (1989). Amino acid geochronology: Resolution and precision in carbonate fossils. *Quat. Int.*, 1: 111–128.

Miller, G.H. and Hare, P.E. (1980). Amino acid geochronology: Integrity of the carbonate matrix and potential of mollusc shells. In: *Biogeochemistry of Amino Acids* (eds. P.E. Hare, T.C. Hoering and K. King Jr.), 415–443. New York: Wiley.

Miller, G.H., Hollin, J.T., and Andrews, J.T. (1979). Aminostratigraphy of UK Pleistocene deposits. *Nature*, 281: 539–543.

Miller, G.H., Magee, J.W., and Jull, A.J.T. (1997). Low-latitude glacial cooling in the Southern Hemisphere from amino acid racemization in emu eggshells. *Nature*, 385: 241-244.

Miller, G.H., Hart, C.P., Roark, E.B., and Johnson, B.J. (2000). Isoleucine epimerization in eggshells of the flightless Australian birds Genyomis and Dromaius. In: *Perspectives in Amino Acid and Protein Geochemistry*, 161–181. Oxford University Press.

Miller, G., Magee, J., Smith, M. et al. (2016). Human predation contributed to the extinction of the Australian megafaunal bird *Genyornis newtoni* ~47 ka. *Nat. Commun.*, 7: 10496.

Müller, P.J. (1984). Isoleucine epimerization in Quaternary planktonic foraminifera; effects of diagenetic hydrolysis and leaching, and Atlantic–Pacific intercore correlations. *Meteor Forschungsergeb., Reihe C*, 38: 25–47.

Müller, P., Staudigel, P.T., Murray, S.T. et al. (2017). Prehistoric cooking versus accurate palaeotemperature records in shell midden constituents. *Sci. Rep.*, 7: 3555.

Murray-Wallace, C.V. and Woodroffe, C.D. (2014). *Quaternary Sea-Level Changes: A Global Perspective*. Cambridge University Press.

New, E., Yanes, Y., Cameron, R.A.D. et al. (2019). Aminochronology and time averaging of Quaternary land snail assemblages from colluvial deposits in the Madeira Archipelago, Portugal. *Quat. Res.*, 92: 483–496.

Oakley, D.O.S., Kaufman, D.S., Gardner, T.W. et al. (2017). Quaternary marine terrace chronology, North Canterbury, New Zealand, using amino acid racemization and infrared-stimulated luminescence. *Quat. Res.*, 87: 151–167.

Oches, E.A., McCoy, W.D., and Clark, P.U. (1996). Amino acid estimates of latitudinal temperature gradients and geochronology of loess deposition during the last glaciation, Mississippi Valley, United States. *Geol. Soc. Am. Bull.*, 108: 892–903.

Olszewski, T.D. and Kaufman, D.S. (2015). Tracing burial history and sediment recycling in a shallow estuarine setting (Copano Bay, Texas) using postmortem ages of the bivalve *Mulinia lateralis. Palaios*, 30: 224–237.

Peltier, W.R. (1998). Postglacial variations in the level of the sea: Implications for climate dynamics and solid-earth geophysics. *Rev. Geophys.*, 36: 603–689.

Penkman, K.E.H., Preece, R.C., Keen, D.H. et al. (2007). Testing the aminostratigraphy of fluvial archives: The evidence from intra-crystalline proteins within freshwater shells. *Quat. Sci. Rev.*, 26: 2958–2969.

Penkman, K.E.H., Kaufman, D.S., Maddy, D., and Collins, M.J. (2008). Closed-system behaviour of the intra-crystalline fraction of amino acids in mollusc shells. *Quat. Geochronol.*, 3: 2–25.

Penkman, K.E.H., Preece, R.C., Bridgland, D.R. et al. (2011). A chronological framework for the British Quaternary based on *Bithynia opercula. Nature*, 476: 446–449.

Penkman, K.E.H., Preece, R.C., Bridgland, D.R. et al. (2013). An aminostratigraphy for the British Quaternary based on *Bithynia opercula. Quat. Sci. Rev.*, 61: 111–134.

Pirazzoli, P.A., Reyss, J.L., Fontugne, M. et al. (2004). Quaternary coral-reef terraces from Kish and Qeshm Islands, Persian Gulf: New radiometric ages and tectonic implications. *Quat. Int.*, 120: 15–27.

Powell, J., Collins, M.J., Cussens, J. et al. (2013). Results from an amino acid racemization inter-laboratory proficiency study; design and performance evaluation. *Quat. Geochronol.*, 16: 183–197.

Ramsey, C.B. (2009). Bayesian analysis of radiocarbon dates. *Radiocarbon*, 51: 337–360.

Rink, W.J. and Thompson, J.W. (2015). *Encyclopedia of Scientific Dating Methods.* Springer.

Ritter, M., Erthal, F., Kosnik, M.A., and Coimbra, J.C. (2017). Spatial variation in the temporal resolution of subtropical shallow-water molluscan death assemblages. *Palaios*, 32: 572–583.

Rutter, N.W., Radtke, U., and Schnack, E.J. (1990). A comparison of ESR and amino acid data in correlating and dating quaternary shorelines along the Patagonian coast. *Argentina. J. Coast. Res.*, 6: 391–411.

Salamon, M., Tuross, N., Arensburg, B., and Weiner, S. (2005). Relatively well preserved DNA is present in the crystal aggregates of fossil bones. *Proc. Natl. Acad. Sci. U.S.A.*, 102: 13783–13788.

Schroeder, R.A. and Bada, J.L. (1976). A review of the geochemical applications of the amino acid racemization reaction. *Earth-Sci. Rev.*, 12: 347–391.

Simonson, A.E., Lockwood, R., and Wehmiller, J.F. (2013). Three approaches to radiocarbon calibration of amino acid racemization in *Mulinia lateralis* from the Holocene of the Chesapeake Bay, USA. *Quat. Geochronol.*, 16: 62–72.

Slezak, M. (2015). Key moments in human evolution were shaped by changing climate. *New Scientist.*, 3039 https://www.newscientist.com/article/mg22730394-100-key-moments-in-human-evolution-were-shaped-by-changing-climate/.

Sykes, G.A., Collins, M.J., and Walton, D.I. (1995). The significance of a geochemically isolated intracrystalline organic fraction within biominerals. *Org. Geochem.*, 23: 1059–1065.

Taylor, R.E., Payen, L.A., Gerow, B. et al. (1983). Middle Holocene age of the Sunnyvale human skeleton. *Science*, 220: 1271–1273.

Taylor, R.E., Payen, L.A., Prior, C.A. et al. (1985). Major revisions in the Pleistocene age assignments for North American human skeletons by C-14 accelerator mass spectrometry: None older than 11,000 C-14 years BP. *Am. Antiq.*, 50: 136–140.

Tomašových, A., Schlögl, J., Kaufman, D.S., and Hudáčková, N. (2016). Temporal and bathymetric resolution of nautiloid death assemblages in stratigraphically condensed oozes (New Caledonia). *Terra Nova*, 28: 271–278.

Tomašových, A., Gallmetzer, I., Haselmair, A. et al. (2017). Stratigraphic unmixing reveals repeated hypoxia events over the past 500 yr in the northern Adriatic Sea. *Geology*, 45: 363–366.

Tomašových, A., Gallmetzer, I., Haselmair, A. et al. (2018). Tracing the effects of eutrophication on molluscan communities in sediment cores: Outbreaks of an opportunistic species coincide with reduced bioturbation and high frequency of hypoxia in the Adriatic Sea. *Paleobiology*, 44: 575–602.

Tomašových, A., Gallmetzer, I., Haselmair, A. et al. (2019). A decline in molluscan carbonate production driven by the loss of vegetated habitats encoded in the Holocene sedimentary record of the Gulf of Trieste. *Sedimentology*, 66: 781–807.

Tomiak, P.J., Penkman, K.E.H., Hendy, E.J. et al. (2013). Testing the limitations of artificial protein degradation kinetics using known-age massive Porites coral skeletons. *Quat. Geochronol.*, 16: 87–109.

Torres, T., Ortiz, J.E., and Arribas, I. (2013). Variations in racemization/epimerization ratios and amino acid content of *Glycymeris* shells in raised marine deposits in the Mediterranean. *Quat. Geochronol.*, 16: 35–49.

Umhoefer, P.J., Maloney, S.J., Buchanan, B. et al. (2014). Late Quaternary faulting history of the Carrizal and related faults, La Paz region, Baja California Sur, Mexico. *Geosphere*, 10: 476–504.

van Duin, A.C.T. and Collins, M.J. (1998). The effects of conformational constraints on aspartic acid racemization. *Org. Geochem.*, 29: 1227–1232.

Vitha, M.F. (2016). *Chromatography: Principles and Instrumentation*. John Wiley & Sons.

Walker, M. (2005). *Quaternary Dating Methods*. Chichester: John Wiley & Sons.

Wehmiller, J.F. (1977). Amino acid studies of the Del Mar, California, midden site: Apparent rate constants, ground temperature models, and chronological implications. *Earth Planet. Sci. Lett.*, 37: 184–196.

Wehmiller, J.F. (1980). Intergeneric differences in apparent racemization kinetics in mollusks and foraminifera: Implications for models of diagenetic racemization. In: *Biogeochemistry of Amino Acids* (eds. P.E. Hare, T.C. Hoering and K. King Jr.), 341–345. New York: Wiley.

Wehmiller, J.F. (1982). A review of amino acid racemization studies in Quaternary mollusks: Stratigraphic and chronologic applications in coastal and interglacial sites, Pacific and Atlantic coasts, United States, United Kingdom, Baffin Island, and tropical islands. *Quat. Sci. Rev.*, 1: 83–120.

Wehmiller, J.F. (1993). Applications of organic geochemistry for quaternary research: Aminostratigraphy and aminochronology. In: *Organic Geochemistry* (eds. M.H. Engle and S.A. Macko), 755–783. New York: Plenum Press.

Wehmiller, J.F. (2013a). United States Quaternary coastal sequences and molluscan racemization geochronology – What have they meant for each other over the past 45 years? *Quat. Geochronol.*, 16: 3–20.

Wehmiller, J.F. (2013b). Interlaboratory comparison of amino acid enantiomeric ratios in Pleistocene fossils. *Quat. Geochronol.*, 16: 173–182.

Wehmiller, J.F. (2015a). Amino acid racemization, coastal sediments. In: *Encyclopedia of Scientific Dating Methods*, 28-32. Springer.

Wehmiller, J.F. (2015b). Amino acid racemization, fluvial and lacustrine sediments (AAR). In: *Encyclopedia of Scientific Dating Methods* (eds. W.J. Rink and J. Thompson), 40-43. Springer.

Wehmiller, J.F. and Belknap, D.F. (1978). Alternative kinetic models for the interpretation of amino acid enantiomeric ratios in Pleistocene mollusks: Examples from California, Washington, and Florida. *Quat. Res.*, 9: 330–348.

Wehmiller, J. and Hare, P.E. (1971). Racemization of amino acids in marine sediments. *Science*, 173: 907–911.

Wehmiller, J.F. and Miller, G.H. (2000). Aminostratigraphic dating methods in Quaternary geology. In: *Quaternary Geochronology*, 187–222. American Geophysical Union.

Wehmiller, J.F. and Pellerito, V. (2015). An evolving database for Quaternary aminostratigraphy. *GeoResJ*, 6: 115–123.

Wehmiller, J.F., Belknap, D.F., Boutin, B.S. et al. (1988). A review of the aminostratigraphy of Quaternary mollusks from United States Atlantic Coastal Plain sites. *Geol. Soc. Am. Special Paper*, 227: 69–110.

Wehmiller, J.F., York, L.L., and Bart, M.L. (1995). Amino acid racemization geochronology of reworked Quaternary mollusks on U.S. Atlantic coast beaches: Implications for chronostratigraphy, taphonomy, and coastal sediment transport. *Mar. Geol.*, 124: 303–337.

Wehmiller, J.F., Stecher, H.A. III, York, L.L., and Friedman, I. (2000). 17, The thermal environment of fossils: Effective ground temperatures at aminostratigraphic sites on the US Atlantic Coastal Plain. In: *Perspectives in Amino Acid and Protein Geochemistry*, 219–251. Oxford University Press.

Wehmiller, J.F., Thieler, E.R., Miller, D. et al. (2010). Aminostratigraphy of surface and subsurface Quaternary sediments, North Carolina coastal plain, USA. *Quat. Geochronol.*, 5: 459–492.

Wehmiller, J.F., Harris, W.B., Boutin, B.S., and Farrell, K.M. (2012). Calibration of amino acid racemization (AAR) kinetics in United States Mid-Atlantic Coastal Plain Quaternary mollusks using $^{87}Sr/^{86}Sr$ analyses: Evaluation of kinetic models and estimation of regional late Pleistocene temperature history. *Quat. Geochronol.*, 7: 21–36.

Wehmiller, J.F., York, L., Pellerito, V. and Thieler, E.R. (2015). Racemization-inferred age distribution of mollusks in the US Atlantic margin coastal system. *Geol. Soc. Am. Meeting*, Paper No. 39-2.

Whiteacre, K., Kaufman, D., Kosnik, M.A., and Hearty, P. (2017). Converting A/I values (ion exchange) to D/L values (reverse phase) for amino acid geochronology. *Quat. Geochronol.*, 37: 1–6.

Wood, R.E., Barroso-Ruíz, C., Caparrós, M. et al. (2013). Radiocarbon dating casts doubt on the late chronology of the Middle to Upper Palaeolithic transition in southern Iberia. *Proc. Natl. Acad. Sci. U.S.A.*, 110: 2781–2786.

Yanes, Y., Kowalewski, M., Ortiz, J.E. et al. (2007). Scale and structure of time-averaging (age mixing) in terrestrial gastropod assemblages from Quaternary eolian deposits of the eastern Canary Islands. *Palaeogeogr. Palaeoclimatol. Palaeoecol.*, 251: 283–299.

Zwartz, D., Bird, M., Stone, J., and Lambeck, K. (1998). Holocene sea-level change and ice sheet history in the Vestfold Hills, East Antarctica. *Earth Planet. Sci. Lett.*, 155: 131–145.

5

Ancient Protein Sequences

5.1 Ancient Protein Analysis by Mass Spectrometry

The field of ancient protein (proteome) analysis using soft-ionization mass spectrometry, or palaeoproteomics, is receiving increasing (and unprecedented) attention. In the past two decades, we have rapidly gone from the first detection of ancient peptides by MALDI-MS (Ostrom et al., 2000; Hollemeyer et al., 2008; Buckley et al., 2009), to the in-depth characterization of ancient proteomes (Cappellini et al., 2012; Warinner et al., 2014a; Demarchi et al., 2016; Sawafuji et al., 2017). A recent application of palaeoproteomics to a human mandible led to the discovery of the first likely Denisovan fossil outside Denisova cave (Chen et al., 2019) and caused some 'hype' within both popular and scientific media (Arnaud, 2019; Warren, 2019). The reader should note, however, that this chapter does not aim to give a detailed overview of the recent advances and application of mass spectrometry-based techniques to ancient proteins; the interested reader is instead directed towards a number of recent reviews (Buckley, 2018; Cappellini et al., 2018; Cleland & Schroeter, 2018; Hendy et al., 2018a; Welker, 2018). Instead, here we will briefly highlight some of the main principles and practical aspects, with a focus on biomineralized samples.

5.1.1 Sample Preparation for Bottom-up Palaeoproteomics

The typical sample preparation workflow for bottom-up proteomics (which is by far the most frequent approach, although top-down studies are also being developed) of ancient biomineralized samples involves the following (Figure 5.1):

Amino acids and Proteins in Fossil Biominerals: An Introduction for Archaeologists and Palaeontologists, First Edition. Beatrice Demarchi.
© 2020 John Wiley & Sons Ltd. Published 2020 by John Wiley & Sons Ltd.

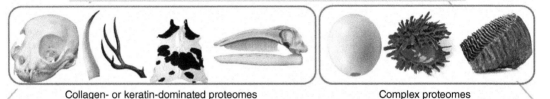

MASS SPECTROMETRY
MALDI-TOF-MS or LC-MS/MS

PURIFICATION AND CONCENTRATION
Solid phase extraction

DIGESTION Trypsin and other proteolytic enzymes

REDUCTION and alkylation of disulphide bridges

EXTRACTION Demineralization, soaking in buffer, lysis, denaturation, gelatinization, EVA/triboelectric

Collagen- or keratin-dominated proteomes   Complex proteomes

Figure 5.1 Schematic of the workflow for bottom-up palaeoproteomics.

- subsampling 1–100 mg of material (this depends on the concentration of proteins within the biomineral, which is typically high for bone, fairly low for eggshell and extremely low in mollusc shell, see Chapter 3); alternatively, a nondestructive method could be used, typically for nonmineralized substrates but also for bone (Fiddyment et al., 2015; Manfredi et al., 2017; McGrath et al., 2019);
- bleaching (if the biomineral is known to retain a closed system);
- demineralization in dilute acid (e.g. 0.6 M HCl, 10% acetic acid) and, where appropriate (e.g. 'young' bone), gelatinization;
- buffer exchange and/or lyophilization and resuspension in a buffer solution (e.g. ammonium bicarbonate) at pH ~7.5;
- reduction and alkylation of disulfide bonds;
- proteolysis using enzymes that cut the sequence at specific amino acid residues, for example chymotrypsin, trypsin, elastase, AspN, collagenase; the choice will depend on the sequence of the protein itself, if known (the aim is to obtain peptides that are 10–20 residues long as this size is optimal for mass spectrometry analysis);
- solid-phase extraction (zip-tip) or other form of sample purification and concentration prior to analysis.

5.1.2   Mass Spectrometry

The study of a proteome using high-resolution liquid chromatography tandem mass spectrometry (typically, nanoLC-MS/MS) is useful for carrying out in-depth protein characterization, including identification, quantification and study of modifications. The sample is ionized, usually by electrospray ionization (ESI), one of the most robust soft-ionization techniques developed in the past few decades (Ho et al., 2003). The ionized peptides (carrying one or multiple charges) are accelerated, selected on the basis of their mass-to-charge ratio (m/z) and these precursor ions are then fragmented further (generally in the gas phase, by CID, collision-induced dissociation, or higher-energy collision-induced dissociation, HCD). The mass spectra of the fragments thus generated (Figure 5.2A), or product ion spectra, can be used to reconstruct the peptide sequence. This operation is called de novo sequencing. A useful guide can be found at: https://www.ionsource.com/tutorial/DeNovo/DeNovoTOC.htm.

Mass analysers using orbitrap technology are favoured as they guarantee excellent mass accuracy, crucial for authentication of ancient sequences. Typically, state-of-the-art instruments for biological applications couple orbitraps with other mass analysers (e.g. ion traps, quadrupoles) in order to maximize flexibility (Figure 5.3). The associated analytical costs are high (~€150 per analysis) and typical analysis times are >1 hour per sample (excluding replicate and blank runs). Bioinformatics is another crucial bottleneck: each analytical run will produce a large file of spectral data, i.e. containing all the numerical values of m/z and intensity values for all product ion spectra. For example, an archaeological eggshell sample will yield around 20 000 product ion spectra, and each of these needs to be interpreted in order to derive the peptide sequence (de novo) and to match this sequence to a database that may contain >200 000 protein sequences of known species. Manual interpretation would be impossible; therefore, specialist software has been developed. Among the most popular search engines are Mascot (www.matrixscience.com), PEAKS Studio (www.bioinfor.com/peaks-studio/) and maxQuant (www.biochem.mpg.de/5111795/maxquant).

Taking into account the high costs and long analytical times, high-resolution palaeoproteomics is mainly suitable for in-depth characterization of fossil proteomes and metaproteomes, identifying ancient disease or food types (Warinner et al., 2014a, 2014b; Hendy, 2015; Hendy et al., 2018b), reconstructing phylogenies of extinct and extant organisms (Cappellini et al., 2012, 2014, 2019; Welker et al., 2015a, 2017; Demarchi et al., 2019; Presslee et al., 2019), or studying the patterns of degradation of proteins, e.g. diagenesis-induced modifications (Wadsworth & Buckley, 2014, 2018; Procopio et al., 2017; Wadsworth et al., 2017; Mackie et al., 2018).

However, high-resolution LC-MS/MS analysis cannot be considered the ideal approach for processing thousands of samples, though this is what is needed, for example, for animal remains from palaeontological and archaeological sites. In order to respond to the archaeologist's need for a robust and cost-effective technique for species identification en masse, Buckley, Collins and colleagues developed the ZooMS (zooarchaeology by mass spectrometry) approach from 2009 onwards (Buckley et al., 2009). ZooMS is in fact a catchy nickname for the well-established technique of

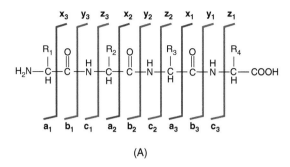

(A)

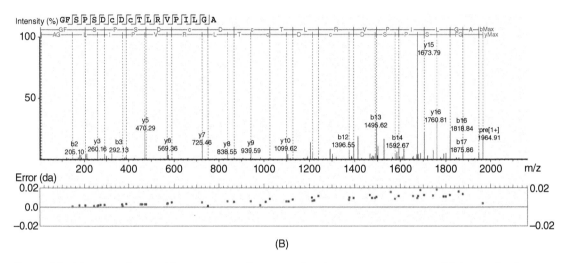

(B)

Figure 5.2 (A) Nomenclature of fragments typically generated in the mass spectrometer and (B) example of a peptide sequence from the molluscan protein Hic74 (derived by assisted de novo using the software PEAKS Studio, Ma et al., 2003).

peptide mass fingerprinting (PMF), first applied to archaeological material (Oetzi's clothing) by Hollemeyer and colleagues in 2008 (Hollemeyer et al., 2008). Despite some variations in the use of the term 'ZooMS' in the early literature, it has now been established within the community that it should be specifically used to indicate the analysis of collagen type I peptides by high-throughput mass spectrometry, typically MALDI-TOF (Figure 5.3). A short video explaining the ZooMS technique can be found here: https:// youtu.be/xBAXaLvGe5I. PMF (and ZooMS) have been used to identify animal (or human) species from fragmentary bone, teeth, antler, ivory, baleen and eggshell (for phylogenetic analyses) (Buckley et al., 2009, 2014, 2017; Rybczynski et al., 2013; von Holstein et al., 2014; Welker et al., 2015a, 2015b, 2017; Brown et al., 2016; Coutu et al., 2016; Demarchi et al., 2016; Prendergast et al., 2017; Solazzo et al., 2017; Presslee et al., 2018), but also textiles, parchment, leather, paintings, organic residues on pots and manufacts (Colonese et al., 2017; Tokarski et al., 2006; Solazzo et al., 2008, 2011; Brandt et al., 2014; Calvano et al., 2015; Fiddyment et al., 2015; Dallongeville et al., 2016; Hendy et al., 2018b).

(A) Bruker Ultraflex MALDI TOF/TOF architecture

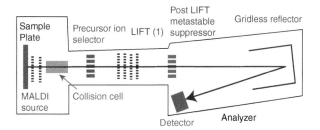

(1) proprietary trademarked term of Bruker Daltonics
used with their TOF-TOF mass spectrometer to describe the
process of elevating the potential of the collision cell above that of the ion source

(B) Thermo Tribrid Fusion architecture

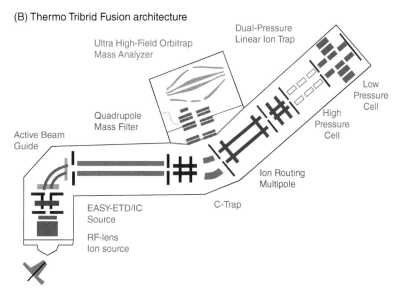

Figure 5.3 (A) Architecture of the Bruker Ultraflex MALDI-TOF/TOF instrument and (B) of the high-resolution Thermo Tribrid Orbitrap Fusion instrument. Adapted and simplified from Suckau et al. (2003) and https://planetorbitrap.com/orbitrap-fusion.

5.1.3 Invertebrate Palaeoproteomics

Invertebrate calcified tissues represent the majority of the organisms listed by Lowenstam (see Table 1.1), and their evolutionary origin is in deep time (see Chapter 1). Indeed, most of the applications of protein diagenesis based on bulk chiral amino acid analyses, from Quaternary geochronology to taxonomy, have focused on invertebrates (Chapter 4). However, palaeoproteomics on invertebrates has so far been attempted in only one study

(Sakalauskaite et al., 2019). In fact, these substrates are especially challenging due to the low concentration of organics trapped within the mineral matrix – as an example, mollusc shells typically contain ~0.5–1% organics by weight, versus ~30% of organics in bone. This is further compounded by the necessity of isolating the stable intracrystalline fraction by bleaching, and thus removing 80–90% of the already minuscule amount of organic material. Low protein concentrations are coupled with small sample sizes – this is the case for both foraminifera and small worked ornaments held in museums – and with variable and complex proteomes. Indeed, as we have seen in Chapter 3, just how complex these proteomes might be is still very much unknown, due to the lack of genomic references and functional studies which would otherwise shed light on the workings of the calcifying matrix.

As a consequence, the typical approach of sample preparation employed by 'classic' palaeoproteomics on protein-rich substrates may not be applicable to invertebrates. This is certainly the case for mollusc shell: proteomic analyses of the calcifying matrix from *modern* shells are typically performed using large amounts of starting material (shell powders) – around 10 g or more. These powders are bleached, decalcified and then thoroughly purified and desalted using ultrafiltration and dialysis (Marie et al., 2010, 2011; Ramos-Silva et al., 2013). However, the sample size for archaeological specimens might be 25–50 mg at best. In order to overcome this, it is necessary to concentrate the proteins, e.g. using ultrafiltration with low cut-off points (otherwise the peptides and short protein strands generated by diagenesis would be lost).

From the analytical point of view, both MALDI-MS and high-resolution LC-MS/MS instruments can be used, with different results and different purposes (Figure 5.3). For example, Figure 5.4 shows MALDI-MS spectra of the intracrystalline proteins isolated from four modern molluscan shell taxa that are commonly recovered from the archaeological record. These preliminary data showed that the intracrystalline peptide fingerprint is different for the four genera (peaks falling at different m/z values), and this implies that the 'ZooMS' approach (i.e. cheap and fast identification by MALDI-MS) can be extended to archaeological mollusc shells. However, we have no sense of the intra-order, intra-family and intra-genus variability, and therefore developing this approach in earnest will entail analysing thousands of shell species in the future.

High-resolution tandem mass spectrometry is more useful than MALDI-MS when conducting in-depth characterization of intracrystalline shell proteomes ('shellomes'): an archaeological Unionoid sample (25 mg) typically yields ~28 000 tandem mass spectra, which can be de novo sequenced by the software and result in ~5000 potential peptide sequences, a number that is comparable to that obtained for other, better characterized, systems, such as bone. Therefore, shellome characterization is theoretically feasible from an analytical point of view.

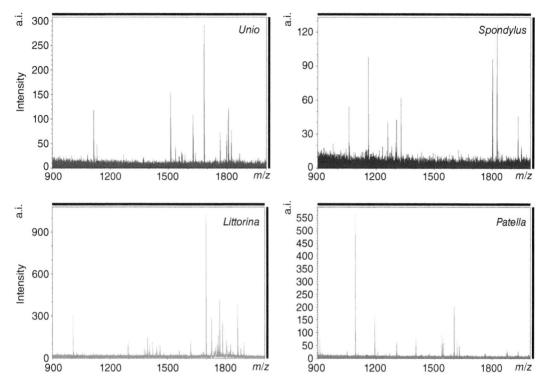

Figure 5.4. MALDI-TOF spectra obtained on the intracrystalline protein fraction isolated from modern *Unio pictorum*, *Spondylus gaederopus*, *Littorina obtusata* and *Patella vulgata*.

The major hurdle to the characterization of the shellome is the bioinformatic step: when these ~5000 peptides are searched against a database of molluscan proteins, only ~1000 actually produce a match with known sequences, the remaining ~4000 being unexplained. Protein identification is highly database-dependent and the quality of the interpretation is proportional to the size of the data set: only about 30 genomes and transcriptomes are available for molluscs, and the phylum contains in the typeset version this is split between lines - shouldn't be~100 000 species. Furthermore, the sequence of known molluscan proteins does not yield a simple phylogenetic signal: it has become clear that a large proportion of shell matrix proteins exhibits long stretches of repeated residues, the RLCDs/LCDs (repetitive low-complexity domains, see Chapter 3) (Marie et al., 2013). Because these RLCDs play a structural role, they are ubiquitous and may be detected in phylogenetically distant lineages (Figure 5.5). This feature is likely shared by proteins from the mineralizing matrix of many organisms, and while it offers the opportunity of discovering which protein domains were important from an evolutionary point of view, it also complicates matters for taxonomic applications. Overall, it is clear that the future of "palaeoshellomics" will be bright once a significant effort is made in characterizing a diverse range of model systems.

5.2 Ancient Proteins: Past and Future

It is interesting to note that the explosion in palaeoproteomics research in the past ten years appears to suggest that, before mass spectrometry, there was nothing. This could not be farther from the truth, as has been discussed throughout this book. A second consideration is that the big open question in palaeoproteomics today is the very same question that has remained unanswered for 60 years: how far back in time can we go when retrieving palaeogenetic information from ancient proteins? This question also brings with it the corollary: where do we find well-preserved proteins? In order to address these questions, it is useful to revisit some of the early work on protein diagenesis and to extrapolate some key findings. It is also worth remembering that, by the time of the discovery of amino acids in fossils, Abelson and his colleagues were already contemplating the idea of fossil protein sequences. In the words of Abelson (1954):

Amino acids have been discovered in association with a variety of invertebrate and vertebrate fossils. An Ordovician trilobite Calymene (approximately 360 million yr old) contained amino acids, three of which have been isolated and identified as alanine, glutamic acid, and valine. These three compounds were also found in a fossil vertebra of the Jurassic dinosaur Stegosaurus. Eight other invertebrate and four vertebrate fossils also contained identifiable amino acids.

The new results on proteins show that peptide bonds may endure for as much as a million years even in a moist environment. Under special conditions where water is not present, peptides would probably be stable for even longer periods.

Biominerals (in which proteins are trapped without much access to water, as we now know from research on intracrystalline closed systems) were of course the preferred substrate from the very beginning (Abelson, 1954).

Particular attention was devoted to $CaCO_3$ fossils. Recent shells of clams, oysters, and tropical snails had 0.1 to 0.2 percent protein made up of most of the common amino acids. Older shells had an amino acid content of 0.1 percent and less. The less stable amino acids, such as tyrosine, were notably diminished or absent in the more ancient remains. The fossil observations have been correlated with laboratory studies of the temperature stability of amino acids. Experiments conducted at 200–260 °C on aqueous solutions and on amino acids contained in the proteins of clam shells have led to additional knowledge of the thermal stability of amino acids. Half of the alanine present in a dilute solution disappeared in 4 hours at 250 °C in a process involving a first-order reaction. From this type of observation, it was estimated that, in storage at room temperature, half of any original alanine could remain after 2 billion yr. These findings have significance for palaeontology and geology.

Soon after, scientists began trying to establish not only a link between macroscopic and microscopic preservation of fossils and preservation of biomolecules, but also the environmental conditions most suited to molecular survival. Of note is the work by Wykoff on collagen survival in dinosaur bones and teeth. His findings, summarized in a paper entitled 'Collagen in fossil bones' (Wyckoff, 1980), include:

- mineral replacement in fossil bones can mimic very accurately minute organic structures;
- preserved collagen (exhibiting the 640 Å spacing characteristic of collagen fibres and detected as constituent amino acids) could be recovered from bones from La Brea tar pits and other Pleistocene samples;
- out of hundreds of Jurassic and Cretaceous bones, only a few contained hydroxyproline (marker of collagen) and instead most contained Ala, Val and Glx (and little Gly);
- preserved amino acids were most likely to be found in bones from alkaline environments;
- the authenticity of these amino acids could not be proved using D/L estimates.

Another important point addressed by researchers in the 1960s and 1970s is that the physical preservation of soft tissues does not imply that the tissues are preserved chemically unaltered. Protein preservation only occurs if there is a mechanism in place that allows it. So far, however, no mechanism has been put forward which can convincingly explain the preservation of peptide bonds over palaeontological timescales (San Antonio et al., 2011). The case of ostrich eggshell peptides surviving in a closed system (dry) for ~4 Ma at the tropics due to their binding to the calcite (causing a thermodynamic 'cooling') highlights the fact that, even under such exceptional conditions, only a few mineral-bound peptides survive (Demarchi et al., 2016). Interestingly, the link between organic matrix ('soft tissue') preservation and protein survival had already been elegantly dismissed by Towe (1980), who analysed the ultrastructure of calcified fossils (the Devonian trilobite arthropod *Phacops rana*) and noncalcified fossils (graptolite periderm) by transmission electron microscopy. Despite the exceptional preservation of the matrix within the trilobite (which is described as 'brownish colored and very delicate, subject to breakage with any but the slightest movement in the EDTA solution'), amino acid analysis failed to detect any residue. Towe concluded that: 'the physical preservation of the organic matrix provides no clues to its chemical degradation, offering instead a misleading criterion for the biogeochemical work we attempted.' He further expounded on this, highlighting that:

1 a mechanism that prevents peptide bond hydrolysis is needed for preservation;
2 an intracrystalline fraction of proteins with limited water is the best bet for finding preserved proteins;

```
  1  MKWCILPLLF GVSAAFLIKC DKRGTSDVSR FEQYVPGHGW VTMWCAVGTG FSPSDCDCTL RVPILGASAA ASAAASAGGF
 81  GGAAAAAAAA AAAAEVVDEF GIGALAGAGA GAGAGGAGAL GLASVLGDVD GLGGLSALGL GGGLAGAGGA DGAGGLSVGG
161  LASLGLGGFG GEGGGGFGGE GEGEGEYEYD DDSSSDSDSD SDENEWDFWN YGSEGGDGGA GFGGGGAGAGA GAGAGVGAGG
241  GGGFGAGAGA GAGAGAGAGA GLGAGIGLGA GLGAGLGAGL GAALGLGGVG GGGLDLDTDT LLALSLGLGI GGGEADDLAL
321  LSLLFGGRQG GSAAAAAAAA AAAGGGGFSG AGAGAGARAN GGIGGFGIGS FGGGIGGGGS AAAAAAAAAA SASRGLGGFG
401  VGAGFDSAAA AAAAASASR GLGGFGVGAG FDAAAAAAAA AAAGASGGGA AAAAAAAAAA AEGQANAAAA AAAAAAAGRG
481  NAAAAASAAA ASIAKKHAAA AAAAAAAAGG AGGSAAAAAA AAAAANRAAA SAAAAAAAVS SGDGSSSASA SASASSSSNG
561  GKRGRGSKDD GSSSATATAS ASASSGGDDK KDSSKLKGLL IKLLESRLSG GASASAAAAA SAGSNGGGRS SSASAAASAS
641  SGDNGKKGSF KDFIKKLLMK VAEKRASGGA SATATASASA SSGGGGFDGG FDGGFDGGFD GGYSAAAAAA AAAAASRRRS
721  YAAAAAAAAA AAGGDGDALA ASAARRRAAA AAAAAAAAAA IRNGQNPIAA AAAAAAAAAS GSGGGGGGNG GGGGSGGSGG
801  GGGAGGSGGS GGSGGSGGSG QRDKVQEEIW KKIDCEEDDN GKYYKRVVKV
```

Figure 5.5. Sequence of protein Hic-74 (originally sequenced from the triangle sail mussel *Hyriopsis cummingii*) identified from an archaeological *Unio crassus* shell (starting sample size: 25 mg, bleached shell powders) by high-resolution LC-MS/MS analysis. Grey peptides have been identified confidently in the archaeological sample; blue underlined peptides are repeated low-complexity domains.

3   even within such systems, the decomposition of hydroxy amino acids will generate some water, eventually driving hydrolysis.

Reassessing this early work is especially important due to the renewed attention of researchers to 'mummified' dinosaur tissues since the advent of palaeoproteomics (Kaye et al., 2008; Manning et al., 2009; Bertazzo et al., 2015). While it is entirely possible that more sensitive mass spectrometric techniques will be able to pick out signals from protein sequences preserved in soft or hard tissues from dinosaurs, it is also true that everything we know (which is, admittedly, not much) about protein breakdown pathways seems to indicate the opposite.

So where does this lead us? There is phylogenetic information locked in ancient proteins, information which we can now partially unlock. Furthermore, because proteins generally degrade at a slower rate, this information takes us beyond the time barrier of ancient DNA. It is thus currently theoretically possible to answer outstanding questions about animal and human evolution, as long as we are within the Pliocene/Pleistocene time frame, and preferably in high-latitude areas. It is likely that new research will lead us to discover new substrates and new analytical approaches able to extend molecular investigations into the deep past.

References

Abelson, P.H. (1954). Amino acids in fossils. *Science,* 119: 576.

Arnaud, C.H. (2019). Ancient proteins tell tales of our ancestors. *Chem. Eng. News,* https://cen.acs.org/analytical-chemistry/art-&-artifacts/Ancient-proteins-tell-tales-ancestors/97/i20 (accessed 19 August 2019).

Bertazzo, S., Maidment, S.C.R., Kallepitis, C. et al. (2015). Fibres and cellular structures preserved in 75-million-year-old dinosaur specimens. *Nat. Commun,* 6: 7352.

Brandt, L.Ø., Schmidt, A.L., Mannering, U. et al. (2014). Species identification of archaeological skin objects from Danish bogs: Comparison between mass spectrometry-based peptide sequencing and microscopy-based methods. *PLoS One*, 9: e106875.

Brown, S., Higham, T., Slon, V. et al. (2016). Identification of a new hominin bone from Denisova Cave, Siberia using collagen fingerprinting and mitochondrial DNA analysis. *Sci. Rep.*, 6: 23559.

Buckley, M. (2018). Zooarchaeology by mass spectrometry (ZooMS) collagen fingerprinting for the species identification of archaeological bone fragments. In: *Zooarchaeology in Practice*, 227–247. Springer.

Buckley, M., Collins, M., Thomas-Oates, J. and Wilson, J.C. (2009). Species identification by analysis of bone collagen using matrix-assisted laser desorption/ionisation time-of-flight mass spectrometry. *Rapid Commun. Mass Spectrom.*, 23: 3843–3854.

Buckley, M., Fraser, S., Herman, J. et al. (2014). Species identification of archaeological marine mammals using collagen fingerprinting. *J. Archaeol. Sci.*, 41: 631–641.

Buckley, M., Harvey, V.L. and Chamberlain, A.T. (2017). Species identification and decay assessment of Late Pleistocene fragmentary vertebrate remains from Pin Hole Cave (Creswell Crags, UK) using collagen fingerprinting. *Boreas*, 46: 402–411.

Calvano, C.D., van der Werf, I.D., Palmisano, F., and Sabbatini, L. (2015). Identification of lipid- and protein-based binders in paintings by direct on-plate wet chemistry and matrix-assisted laser desorption ionization mass spectrometry. *Anal. Bioanal. Chem.*, 407: 1015–1022.

Cappellini, E., Jensen, L.J., Szklarczyk, D. et al. (2012). Proteomic analysis of a Pleistocene mammoth femur reveals more than one hundred ancient bone proteins. *J. Proteome Res.*, 11: 917–926.

Cappellini, E., Gentry, A., Palkopoulou, E. et al. (2014). Resolution of the type material of the Asian elephant, *Elephas maximus* Linnaeus, 1758 (Proboscidea, Elephantidae). *Zool. J. Linn. Soc.*, 170: 222–232.

Cappellini, E., Prohaska, A., Racimo, F. et al. (2018). Ancient biomolecules and evolutionary inference. *Annu. Rev. Biochem.*, 87: 1029–1060.

Cappellini, E., Welker, F., Pandolfi, L. and Madrigal, J.R. (2019). Early Pleistocene enamel proteome sequences from Dmanisi resolves Stephanorhinus phylogeny. *Nature*, 574: 103–107.

Chen, F., Welker, F., Shen, C.-C. et al. (2019). A late Middle Pleistocene Denisovan mandible from the Tibetan Plateau. *Nature*, 569: 409–412.

Cleland, T.P. and Schroeter, E.R. (2018). A comparison of common mass spectrometry approaches for paleoproteomics. *J. Proteome Res.*, 17: 936–945.

Colonese, A.C., Hendy, J., Lucquin, A. et al. (2017). New criteria for the molecular identification of cereal grains associated with archaeological artefacts. *Sci. Rep.*, 7: 6633.

Coutu, A.N., Whitelaw, G., le Roux, P. and Sealy, J. (2016). Earliest evidence for the ivory trade in Southern Africa: Isotopic and ZooMS analysis of seventh–tenth century AD ivory from KwaZulu-Natal. *Afr. Archaeol. Rev.*, 33: 411–435.

Dallongeville, S., Garnier, N., Rolando, C., and Tokarski, C. (2016). Proteins in art, archaeology, and paleontology: From detection to identification. *Chem. Rev.*, 116: 2–79.

Demarchi, B., Hall, S., Roncal-Herrero, T. et al. (2016). Protein sequences bound to mineral surfaces persist into deep time. *eLife*, 5: e17092.

Demarchi, B., Presslee, S., Gutiérrez-Zugasti, I. et al. (2019). Birds of prey and humans in prehistoric Europe: A view from El Mirón Cave, Cantabria (Spain). *J. Archaeol. Sci.: Rep.*, 24: 244–252.

Fiddyment, S., Holsinger, B., Ruzzier, C. et al. (2015). Animal origin of 13th-century uterine vellum revealed using noninvasive peptide fingerprinting. *Proc. Natl. Acad. Sci. U.S.A.*, 112: 15066–15071.

Hendy, J. (2015). Ancient metaproteomics: A novel approach for understanding disease and diet in the archaeological record. PhD thesis, University of York.

Hendy, J., Welker, F., Demarchi, B. et al. (2018a). A guide to ancient protein studies. *Nat. Ecol. Evol.*, 2: 791–799.

Hendy, J., Colonese, A.C., Franz, I. et al. (2018b). Ancient proteins from ceramic vessels at Çatalhöyük West reveal the hidden cuisine of early farmers. *Nat. Commun.*, 9: 4064.

Ho, C.S., Lam, C.W.K., Chan, M.H.M. et al. (2003). Electrospray ionisation mass spectrometry: Principles and clinical applications. *Clin. Biochem. Rev.*, 24: 3–12.

Hollemeyer, K., Altmeyer, W., Heinzle, E. and Pitra, C. (2008). Species identification of Oetzi's clothing with matrix-assisted laser desorption/ionization time-of-flight mass spectrometry based on peptide pattern similarities of hair digests. *Rapid Commun. Mass Spectrom.*, 22: 2751–2767.

Kaye, T.G., Gaugler, G. and Sawlowicz, Z. (2008). Dinosaurian soft tissues interpreted as bacterial biofilms. *PLoS One*, 3: e2808.

Ma, B., Zhang, K., Hendrie, C. et al. (2003). PEAKS: Powerful software for peptide de novo sequencing by tandem mass spectrometry. *Rapid Commun. Mass Spectrom.*, 17: 2337–2342.

Mackie, M., Rüther, P., Samodova, D. et al. (2018). Palaeoproteomic profiling of conservation layers on a 14th century Italian wall painting. *Angew. Chem. Int. Ed Engl.*, 57: 7369–7374.

Manfredi, M., Barberis, E., Gosetti, F. et al. (2017). Method for noninvasive analysis of proteins and small molecules from ancient objects. *Anal. Chem.*, 89: 3310–3317.

Manning, P.L., Morris, P.M., McMahon, A. et al. (2009). Mineralized soft-tissue structure and chemistry in a mummified hadrosaur from the Hell Creek Formation, North Dakota (USA). *Proc. Biol. Sci.*, 276: 3429–3437.

Marie, B., Marie, A., Jackson, D.J. et al. (2010). Proteomic analysis of the organic matrix of the abalone *Haliotis asinina* calcified shell. *Proteome Sci.*, 8: 54.

Marie, B., Trinkler, N., Zanella-Cleon, I. et al. (2011). Proteomic identification of novel proteins from the calcifying shell matrix of the Manila clam *Venerupis philippinarum*. *Mar. Biotechnol.*, 13: 955–962.

Marie, B., Jackson, D.J. and Ramos-Silva, P. (2013). The shell-forming proteome of *Lottia gigantea* reveals both deep conservations and lineage-specific novelties. *FEBS J.*, 280: 214–232.

McGrath, K., Rowsell, K., Gates St-Pierre, C. et al. (2019). Identifying archaeological bone via non-destructive ZooMS and the materiality of symbolic expression: Examples from Iroquoian bone points. *Sci. Rep.*, 9. 11027.

Ostrom, P.H., Schall, M., Gandhi, H. et al. (2000). New strategies for characterizing ancient proteins using matrix-assisted laser desorption ionization mass spectrometry. *Geochim. Cosmochim. Acta*, 64: 1043–1050.

Prendergast, M.E., Buckley, M., Crowther, A. et al. (2017). Reconstructing Asian faunal introductions to eastern Africa from multi-proxy biomolecular and archaeological datasets. *PLoS One*, 12: e0182565.

Presslee, S., Wilson, J., Russell, D.G.D. et al. (2018). The identification of archaeological eggshell using peptide markers. *STAR: Sci. Technol. Archaeol. Res.*, 3: 89–99.

Presslee, S., Slater, G.J., Pujos, F. et al. (2019). Palaeoproteomics resolves sloth relationships. *Nat. Ecol. Evol.*, 3: 1121–1130.

Procopio, N., Chamberlain, A.T. and Buckley, M. (2017). Intra- and interskeletal proteome variations in fresh and buried bones. *J. Proteome Res.*, 16: 2016–2029.

Ramos-Silva, P., Marin, F., Kaandorp, J., and Marie, B. (2013). Biomineralization toolkit: The importance of sample cleaning prior to the characterization of biomineral proteomes. *Proc. Natl. Acad. Sci. U.S.A.*, 110: E2144–E2146.

Rybczynski, N., Gosse, J.C., Richard Harington, C. et al. (2013). Mid-Pliocene warm-period deposits in the High Arctic yield insight into camel evolution. *Nat. Commun.*, 4: 1550.

Sakalauskaite, J., Andersen, S., Biagi, P. et al. (2019). "Palaeoshellomics" reveals the use of freshwater mother-of-pearl in prehistory. *eLife*, 8: e45644.

San Antonio, J.D., Schweitzer, M.H., Jensen, S.T. et al. (2011). Dinosaur peptides suggest mechanisms of protein survival. *PLoS One*, 6: e20381.

Sawafuji, R., Cappellini, E., Nagaoka, T. et al. (2017). Proteomic profiling of archaeological human bone. *R. Soc. Open Sci.*, 4: 161004.

Solazzo, C., Fitzhugh, W.W., Rolando, C. and Tokarski, C. (2008). Identification of protein remains in archaeological potsherds by proteomics. *Anal. Chem.*, 80: 4590–4597.

Solazzo, C., Heald, S., Ballard, M.W. et al. (2011). Proteomics and Coast Salish blankets: A tale of shaggy dogs? *Antiquity*, 85: 1418–1432.

Solazzo, C., Fitzhugh, W., Kaplan, S., Potter, C. and Dyer, J.M. (2017). Molecular markers in keratins from Mysticeti whales for species identification of baleen in museum and archaeological collections. *PLoS One*, 12: e0183053.

Suckau, D., Resemann, A., Schuerenberg, M. et al. (2003). A novel MALDI LIFT-TOF/TOF mass spectrometer for proteomics. *Anal. Bioanal. Chem.*, 376: 952–965.

Tokarski, C., Martin, E., Rolando, C., and Cren-Olivé, C. (2006). Identification of proteins in renaissance paintings by proteomics. *Anal. Chem.*, 78: 1494–1502.

Towe, K.M. (1980). Preserved organic ultrastructure: An unreliable indicator for Paleozoic amino acid biogeochemistry. In: *Biogeochemistry of Amino Acids*, vol. 558 (eds. P.E. Hare, T.C. Hoering and K. King Jr.). New York: Wiley.

von Holstein, I.C.C., Ashby, S.P., van Doorn, N.L. et al. (2014). Searching for Scandinavians in pre-Viking Scotland: Molecular fingerprinting of Early Medieval combs. *J. Archaeol. Sci.*, 41: 1–6.

Wadsworth, C. and Buckley, M. (2014). Proteome degradation in fossils: Investigating the longevity of protein survival in ancient bone. *Rapid Commun. Mass Spectrom.*, 28: 605–615.

Wadsworth, C. and Buckley, M. (2018). Characterization of proteomes extracted through collagen-based stable isotope and radiocarbon dating methods. *J. Proteome Res.*, 17: 429–439.

Wadsworth, C., Procopio, N., Anderung, C. et al. (2017). Comparing ancient DNA survival and proteome content in 69 archaeological cattle tooth and bone samples from multiple European sites. *J. Proteomics*, 158: 1–8.

Warinner, C., Rodrigues, J.F.M., Vyas, R. et al. (2014a). Pathogens and host immunity in the ancient human oral cavity. *Nat. Genet.*, 46: 336–344.

Warinner, C., Hendy, J., Speller, C. et al. (2014b). Direct evidence of milk consumption from ancient human dental calculus. *Sci. Rep.*, 4: 7104.

Warren, M. (2019). Move over, DNA: Ancient proteins are starting to reveal humanity's history. *Nature*, 570: 433–436.

Welker, F. (2018). Palaeoproteomics for human evolution studies. *Quat. Sci. Rev.,* 190: 137–147.

Welker, F., Collins, M.J., Thomas, J.A., Wadsley, M. and Brace, S. (2015a). Ancient proteins resolve the evolutionary history of Darwin's South American ungulates. *Nature*, 522: 81–84.

Welker, F., Soressi, M., Rendu, W., Hublin, J.-J. and Collins, M. (2015b). Using ZooMS to identify fragmentary bone from the late Middle/Early Upper Palaeolithic sequence of Les Cottes, France. *J. Archaeol. Sci.,* 54: 279–286.

Welker, F., Smith, G.M., Hutson, J.M. et al. (2017). Middle Pleistocene protein sequences from the rhinoceros genus *Stephanorhinus* and the phylogeny of extant and extinct Middle/Late Pleistocene Rhinocerotidae. *PeerJ*, 5: e3033.

Wyckoff, R.W.G. (1980). Collagen in fossil bones. In: *Biogeochemistry of Amino Acids* (eds. P.E. Hare, T.C. Hoering and K. King Jr.), 17–22. New York: Wiley.

Index

amino acid
  bound (peptide-bound)   10, 14, 24,
    26, 29, 76–79, 88, 121
  free (FAA, free amino acids)   14,
    24, 28–29, 32–35, 74, 76–79,
    88, 94, 103
  terminal *vs.* internal   29, 34, 78,
    83, 88
  total (THAA, total hydrolysable
    amino acids)   14, 24, 28–29,
    32–35, 74, 76–79, 88, 94, 103
amino acid racemization (AAR)   3,
    12, 15, 24, 28, 31–35, 76–103
aminochronology   78, 83, 92–97
aminostratigraphy   78, 85, 87–92
antler   45
aragonite, *see* calcium carbonate
Arrhenius, parameters   92–94
arthropods   6, 57, 62–63
authentication   38, 115

Bayesian methods   74–75, 94
biomineral, biomineralization   1–8
  biomineralization, toolkit   5–8
bleaching   5, 13–16, 35, 49, 78, 83,
    101–102, 114, 118
bone   3, 4, 14, 15, 30, 43–45, 74, 83,
    114, 121
  collagen, bone   43–45, 74, 83,
    116, 121
brachiopods   15, 57, 61–62

calcium carbonate   4, 6, 9, 47, 59,
    62–63
  amorphous   2, 4–9, 47, 54, 62–63

aragonite, nacre, mother-of-
    pearl   2, 4, 8, 25, 52–56, 87,
    99, 102
calcite   2, 4, 9, 25, 30–31, 47–50,
    52, 56, 62, 87, 99, 121
calcite, *see* calcium carbonate
calculus, dental   46
calibration   78, 83, 92–96, 103
Cambrian   5
carbohydrates, sugars   3–4, 24, 29,
    54, 58
chirality, chiral amino acids   31–33,
    71–103
chitin   4, 8, 54, 62–63
chromatography   76, 79–82
  gas chromatography (GC)   79
  ion-exchange liquid
    chromatography (IEC)   79
  reverse-phase high-pressure liquid
    chromatography
    (RP-HPLC)   79
  ultra high-pressure liquid
    chromatography
    (UHPLC)   82
closed system   13–16, 25, 28–29, 31,
    35–36, 38, 45, 50, 77–78,
    82–84, 90, 92, 98–99, 101, 114,
    120, 121
condensation   11, 24, 27–28, 30
coral   5, 15, 36, 57–59, 95

deamidation   26, 34–38
decomposition   26, 35–38, 122
dehydration   26, 35–36
demineralization   78, 114

Amino acids and Proteins in Fossil Biominerals: An Introduction for Archaeologists and Palaeontologists, First Edition. Beatrice Demarchi.
© 2020 John Wiley & Sons Ltd. Published 2020 by John Wiley & Sons Ltd.

diagenesis   3, 12, 15–16, 23–27
    experiments (artificial
        diagenesis)   26, 28, 32, 34–35,
        60, 78, 92–94, 101–102
    mechanisms of diagenesis   23–38
diketopiperazine (DKP)   34
DL ratio, D/L value   31–34, 76–103

eggshell   3, 8, 9, 11, 15, 25–31, 36–38,
    47–51, 82–83, 86, 94, 97, 98,
    102, 114–116, 121
enantiomer   31–34, 76–80
endogenous   3
epimerization   32, 76
eustasy (sea level change)   72–73
exogenous   13–14, 49

foraminifera   1, 2, 36, 57, 59–61, 83,
    94, 96, 99, 102
fossil   1, 3, 12, 16, 24, 36, 43, 60, 71,
    76, 79, 120–122

geochronology   16, 44, 50, 57, 59,
    71–103

Holocene   71, 83, 85, 86, 87, 92,
    95–96, 97
horn   45
hydrolysis   24–26, 27–38, 94, 121–122
    acid hydrolysis   29, 34, 76, 78,
        80, 103

intercrystalline   5, 13–15, 54
intracrystalline   5, 13–16, 25, 28, 30,
    31, 35–37, 49, 54, 78, 82, 83, 84,
    85, 88, 98, 99, 101, 118,
    120–121
intracrystalline protein dating
    (IcPD)   89
isostasy (sea level change)   72–73
ivory   46, 99, 115

kinetics   28–29, 33–34, 60, 84, 86,
    92–97, 101
    first order, irreversible   28–29
    first order, reversible   33–34

Last Glacial Maximum (LGM)   50,
    73, 97

Last Interglacial   73, 90, 92
leaching   13–15, 35, 101
lectin, C-type   9, 49–50
lipids   3, 4, 23–24, 54–55, 58
low complexity domains, repeated
    (RLCD)   9, 57

Marine (oxygen) Isotope Stage
    (MIS)   71–72, 74, 77, 85, 89,
    90, 96, 98
mass spectrometry   37, 44, 83, 113–117
    tandem mass spectrometry   55, 61,
        94, 113–117
matrix, organic   4–7, 9, 13–15, 25, 54,
    56, 58, 59, 60, 61, 62, 63, 82,
    118, 119, 121
melanoidin   29
microstructure   8, 13, 25, 30, 48, 52,
    54, 99, 102
mixing, ages, *see* time averaging
mixing, sediments, *see* reworking
modification, amino acid
    diagenesis-induced
        modifications   26, 35–37, 115
    post-translational modifications
        (PTMs)   11, 115
mother-of-pearl, *see* aragonite

nacre, *see* aragonite

open system   13–14, 44, 82–83
ovocleidin   9, 11, 48–50

palaeobiogeochemistry   3, 23, 32
palaeoecology   85, 95–96
palaeoproteomics   3, 15–16, 57, 61,
    113–122
palaeothermometry   50, 60, 97–98
peptide (definition of)   10–11
peptide mass fingerprint (PMF), *see*
    zooarchaeology by mass
    spectrometry (ZooMS)
periostracum   52–54
Pleistocene   44, 71–75, 90, 96,
    121–122
protein (definition of)   10–11
    structure   10–11
    synthesis   27–28
proteome, biomineralized   8–9,
    12–16, 25–26, 43–63, 113–120

Quaternary, period   12, 44, 50, 71–76, 85–86, 90–92, 96–97

racemization   3, 12, 15, 24, 26, 28, 31–35, 36, 44, 50, 60, 73, 75, 76–78, 83, 84, 86, 88, 90, 94, 95, 96, 97, 98, 100, *see also* amino acid racemization (AAR)
radiocarbon   44, 51, 73–75, 78, 83, 86, 92, 95, 96, 97, 99, 101
raised beach   73, 84–85, 89, 97
relative dating, relative age information   12, 77–78, 87–92, *see also* aminostratigraphy
reworking   85–86

sampling   82, 86, 101–102, 114
shell, mollusc   8, 9, 15, 23–25, 29, 30, 49, 51–57, 60, 74, 76, 82, 84, 86, 95, 102, 114, 118–120
    mantle, shell   9, 54–56
    midden, shell   51, 62, 86, 97
shellome   54–57, 118
soft ionization methods   55, 113–117
    electrospray (ESI)   115
    matrix-assisted laser desorption/ionization (MALDI)   94, 113–119

species effect   84, *see also* taxonomy
stereoisomer   31–32, 80
struthiocalcin   *31 (Fig. 2.3)*, 36, *37 (Fig. 2.5)*, 49–50
sub-fossil   3
succinimide   34, 36
sugars, *see* carbohydrates

taxonomy, chemical   57, 60, 61, 84, 99–100, 117–120
temperature   1, 5, 25–26, 29, 34–35, 38, 50–51, 59, 60, 62, 71, 76–78, 86–87, 89, 90, 92–99, 101, 103, 120
    mean annual temperature (MAT)   86
thermal age   86, 98
time averaging   84–85, 95–96
tooth, enamel/dentin   3, 45–46

water   4, 11, 14, 15, 16, 26, 27, 28–31, 37, 47, 50, 59, 72, 101, 102, 120, 121, 122

zooarchaeology by mass spectrometry (ZooMS)   115–119